TURNING THE TIDE ON PLASTIC

Journalist, broadcaster and eco expert Lucy Siegle has had a weekly prime-time TV slot dedicated to battling waste plastic (*The One Show*, BBC1) and a decade of experience as the *Observer* and *Guardian*'s Ethical Living columnist. Lucy founded the Green Carpet Challenge with Livia Firth to address consumption and sustainability in the fashion industry, and recently worked on environmental projects with Emma Watson and Ellie Goulding. She is well known for her enthusiasm, optimism and playful authority. She recently chaired a panel of some of the world's most exciting plastic activists planning a post-plastic future at the third UN Environment Assembly in Kenya

You can follow Lucy on:

Twitter @lucysiegle
Instagram @theseagull

LUCY SIEGLE

TRAPEZE

First published in Great Britain in 2018 by Trapeze,
an imprint of The Orion Publishing Group Ltd
Carmelite House, 50 Victoria Embankment,
London EC4Y 0DZ

An Hachette UK company

1 3 5 7 9 10 8 6 4 2

A CIP catalogue record for this book is
available from the British Library.

ISBN (Hardback): 978 1 409 18298 6
ISBN (eBook): 978 1 409 18300 6

Printed and bound in Britain by
Clays Ltd, Elcograf S.p.A

To my nieces and nephews,
Ava, Heidi, Jasper, Nina, Sam and Zoe.
In the hope that we will one day bring you clean oceans
full of fish and cetaceans, not plastic.

CONTENTS

FOREWORD

I first met Lucy for a beach clean in Cornwall a decade ago, just as I took the helm of Surfers Against Sewage, a charity dedicated to the protection of oceans, waves, beaches and wildlife. We picked up crisp packets, cotton bud sticks, plastic bottles and all sorts of plastic that simply shouldn't have been on the beach. This was long before plastic pollution was making daily headlines, or community-driven plastic campaigns were causing shockwaves across government and industry. Lucy is one of the true plastic pollution pioneers.

It is the beach environment that first exposed the damage caused by humanity's addiction to plastic and the haemorrhaging systems that are generating a tidal wave of plastic pollution: a juxtaposition of excessive and unnecessary consumption set against some of the most beautiful places on earth. Picture-postcard places. Those most connected with our oceans and coastlines – surfers, beach lovers, swimmers, sailors – were the canary in the coalmine. Some of the first to see tidelines of plastic: some of the first to question why it was there. The first to sound the alarm.

But how did it get to this?

Plastic is an extraordinary material – flexible, colourful, light, abundant and almost indestructible. It has had an impact on every human industry and revolutionised the very way we live. It has helped us explore the farthest reaches of the planet, from the deepest ocean trenches to the highest mountain ranges, and even beyond our atmosphere, through space travel.

Plastic is also an extraordinary pollutant. *Flexible, colourful, light, abundant and almost indestructible*: the very properties that make it so useful also make it problematic when it escapes into the environment. The sea is often the final destination of plastic waste, as it is carried by winds, streams, rivers and currents into oceanic whirlpools of pollution. Indeed, it seems increasingly unlikely that a pristine, 'plastic-free' environment today truly exists on any part of Planet Earth, or should I say Planet Ocean. As Jacques Cousteau said, 'Water and air, the two essential fluids on which all life depends, have become global garbage cans.'

However, plastic is also becoming an extraordinary unifier. The qualities which make it so useful and so problematic make it the pollutant that everyone can see, identify and respond to in their everyday lives. The millions of tonnes of plastics that enter our oceans annually are the same as the millions of tonnes that people interact with every day in our shops, restaurants, homes and offices.

The tideline on most beaches usually has the same plastic brand profile as any homogenised high street. Immediately recognisable to anyone taking part in a beach clean, this jetsam is a warning sign of a failing linear economy, excessive packaging, planned obsolescence, big oil interests, poor resource management and fragmented recycling systems.

Scenes of beaches, cities and wilderness choked with plastic have also generated a tidal wave of social action. To quote Jacques Cousteau again, 'The sea, the great unifier, is man's only hope. Now, as never before, the old phrase has a literal meaning: we are all in the same boat.'

The ubiquitous nature of plastic pollution and its relevance to our daily lives is perhaps unique among the most ominous pollutants of our times. Unlike carbon dioxide and other invisible contaminants, we can all visualise it as part of our daily existence.

A global army of beach clean volunteers today tackles the plastic pollution at the frontline, where land and sea converge, interlocked by a swathe of plastic. Removing and recycling as much plastic as possible from our beaches, streets and countryside is important work: every piece removed a victory in its own right, a small battle won. However, we know that we can't simply pick our way out of the problem. We must urgently focus efforts further upstream in the war on avoidable plastics. Plastic in our environment is not litter. It is a pollutant from fast business serving society in the wrong way while outsourcing the cost to Mother Nature.

Our political leaders must take bold, decisive steps to break this industry and society's love affair with single-use plastics. In the UK, the extension of the five-pence plastic bag charge, the microbead ban and a UK-wide deposit refund system can be great steps to reduce the plastic footprint that is trampling our natural world. But these measures are only the start.

Industry transparency and accountability, taking full responsibility for its plastic emissions, is essential to driving more sustainable, circular systems and products. Governments must legislate to incentivise businesses to reconsider their plastic footprint and decouple from pointless plastics, which no one actually wants. Progressive businesses must use plastic-free policies as a market advantage to respond to the growing global community that demands it.

We must reinvent our relationship with single-use plastic to eliminate, replace and recycle plastics faster and more effectively. Plastic production is already rampant and is set to increase massively in the next twenty-five years. And all this avoidable plastic is made from oil: oil that is pumped from the ground beneath oceans, fields, countryside and precious wilderness. Boycotting single-use plastic is also taking a stand against new oil exploration and extraction – a double high-five for the planet.

The time for action is now, and it's up to each of us not to kick the plastic bottle down the street. We can all reduce our plastic footprint, and Lucy and her strategies in this book will show you how. Every day, every week, every consumer choice we make. A consumer revolt against avoidable plastics is already under way. World leaders are responding. Businesses are seeing that they need to change. But we must keep up the pressure. We cannot put the genie of plastic pollution back in the bottle.

Together we can create a plastic-free future.

Hugo Tagholm
CEO, Surfers Against Sewage
Cornwall, 10 May 2018

INTRODUCTION

WELCOME TO THE PLASTIC AGE

The plastic we throw away in a single year could circle the earth four times. Out of the 320 million metric tonnes of new plastic mass-produced each year[1] – almost all from oil – eight million tonnes leak into the world's oceans and waterways. That is the equivalent of a truckload of plastic being upturned and shaken out straight into the sea every minute of every day. Every minute of every day, one million plastic bottles are used.[2] Imagine each of those bottles, a quarter filled with oil; the amount of oil needed to make the bottle in the first place.[3] In the last decade we've produced more plastic than we did during the whole of the last century. And – this is the one that usually stops people in

1. Martinko, K.: '"A Plastic Tide" film depicts shocking plastic pollution worldwide', Treehugger, 26 January 2017.
2. Nac, Trevor: 'We're Now At A Million Plastic Bottles Per Minute – 91% Of Which Are Not Recycled', Forbes.com, 26 July 2017.
3. Clarke Fox, C.: 'Drinking Water: Bottled or From the Tap?', National Geographic Kids, Info graphic, 2016.

their tracks – by 2050 the ocean will contain more plastic by weight than fish.[4]

Plastic has us in a vice-like grip. It has colonised supermarket shelves and kitchen bins; invaded parks, grass verges, beaches and beauty spots. It has leaked into our oceans to impact wildlife and muscled its way on to the nightly news. For a material that's supposed to provide background assistance to everyday life, that's quite the attention upgrade.

Plastic has jostled its way into our very souls, but more alarmingly still, it has been shown to be in both our food chain and our bodies. Cleaning up the unwanted plastic from all of these areas and halting its further march into the fabric of our lives starts right now. It has to. In human history, as many have observed, there was the Stone Age, the Bronze Age, the Iron Age and today we are living through the Plastic Age. But the Plastic Age is not something we can sit by and watch passively, observing as if the inexorable plastic takeover was just another natural phase of human evolution. As I'll show in this book, the impact of the plastic pandemic is so serious, it becomes a zero-sum game. Either plastic wins or we do.

ONE BILLION ELEPHANTS . . .

The last twelve months have unleashed not just another 320 million tonnes of virgin plastic – that's brand-new plastic

4. Ellen MacArthur Foundation.

from newly extracted oil – but an abundance of analysis as to what that may or may not mean for our planet. Perhaps out of all the figures that have been posited, the report that really rocked me was by the well-respected industrial ecologist and academic Professor Roland Geyer, heading up a team of US researchers. This was the first global analysis of *all* the plastic production there has ever been[5] and it blew my socks off.

Professor Geyer's report shows that in total humans have produced 8.3 billion tonnes of plastic since its industrial-scale production really got going in the 1950s. That's the weight of one billion elephants. By 2015 just 9 per cent had been recycled, 12 per cent incinerated and 79 per cent had accumulated in landfills or the wider environment. That means nearly all the plastic that has ever been produced is still with us.

The ecological impact of our plastic production is plain to see. From the Great Barrier Reef, which now requires a $500 million 'rescue' package from the Australian government, to the fast-melting glaciers in Alaska, the planet is sending us clear and insistent distress signals: that we must take urgent action to halt global warming and habitat destruction. All the while, however, we persist in creating yet more – more new plastic, and more plastic waste. Ignoring all of the evidence and these alarms from the natural world, plastic production and consumption continues to rise at a relentless pace.

This is not a war on all plastics. As pointed out in 'The New Plastics Economy', a seminal report from the Ellen

5. Geyer, R., Jambeck, J. R., Lavender Law, K.: 'Production, use, and fate of all plastics ever made', *Science Advances*, 19 Jul 2017: vol. 3, no. 7, e1700782.

MacArthur Foundation, plastics now make up 15 per cent of the average car (making them lighter and more fuel- or electricity-efficient), and approximately 50 per cent of the Boeing Dreamliner. The report notes that 'imagining a world without plastics is nearly impossible'.[6] Given that the aim of the report was to evaluate the potential for an alternative to an increasingly, even entirely plastic planet, this speaks volumes – if the leading minds in innovation can't imagine life without plastics, then that means it's pretty tough to imagine them gone completely.

In any case, this might be neither desirable nor realistic. Some polymers perform critical, life-saving functions, quite literally: plastic forms heart valves and is used in bulletproof vests. Its low weight and incredible heat resistance means it is necessary for space shuttles. From ice rinks to sports pitches and children's playgrounds and Lego, it enables our leisure lives, too. Of course there are polymers with noble ends and heroic applications. That's irrefutable.

What this is, is a battle against avoidable, unwanted, useless, nuisance plastic that is unnecessarily forced upon us and then into the natural environment, where its impact is devastating.

Not a day goes by when I don't hear of a miracle cure for the plastic plague. Don't worry, I'm told, our heaps of discarded plastic will soon be eaten by enzymes deployed outside the local supermarket or blasted off, moon-bound, in a rocket. Respectfully, I disagree with these predictions.

6. www.ellenmacarthurfoundation.org/publications/the-new-plastics-economy-rethinking-the-future-of-plastics-catalysing-action

There are some exciting breakthroughs, but they are at a very early stage. There is no silver bullet. The plastic problem is a lot more complex than the optimists would like us to think.

The stakes are high and the solutions far from simple. Our politicians (of all persuasions) have a shaky relationship with the planet. There is often little appetite at government level for the radical action that's needed to stop an environmental disaster. Obfuscation and confusion rule, as blame and responsibility are shifted between states and governments. The democratic political cycle lasts on average four years, rather than geological epochs that last tens of thousands, so you can see why the political system is not exactly set up to take a long-term sustainable view. Many environmental issues remain an inconvenient truth. It would be more expedient to carry on as usual.

As Professor Geyer put it himself, 'We cannot continue with business as usual unless we want a planet that is literally covered in plastic.'[7] Who on earth would want that?

TURNING THE TIDE ON PLASTIC

Huge numbers of us are now waking up to the plastic pandemic and deciding that if governments and global agencies won't or can't move fast enough, then we are going to have to do something ourselves. But how? What is the most effective form of action we can take?

7. Cohen, J.: 'A Plastic Planet', *The Current*, UC Santa Barbara, CA, 19 July 2017.

Turning the Tide on Plastic is, in essence, a survival guide. I'm quite the fan of this genre. I'm always giving out survival guides as Christmas presents, and I avidly watch shows like *The Island* on Channel 4. This means my brother-in-law knows what to do in the event of meeting a grizzly bear and I know, theoretically, how to kill a caiman for food (I wouldn't. I'm vegetarian). Never mind that we will most likely never face either scenario. Thanks to the popularity of adventurists such as Bear Grylls, millions of people are equipped with an arsenal of handy survival skills – at least in their heads. But isn't it curious that we're so prepared for life on Bear's island, and so ill-prepared for emergencies on our own? Sometimes fighting a threat is particularly difficult simply *because* it's under your nose and in your own home, and so becomes oddly familiar.

That's where this book comes in. In *Turning the Tide on Plastic*, I want to unwrap everything about plastic – from its creation to its likely destination – to equip you with the latest information in order to make up your own mind about the use and misuse of plastic, and give you practical tips and strategies to help you make choices or changes today to help our planet of tomorrow.

Plastic has become an unavoidable fact of modern life. As with anything habitual and longstanding, we need first to understand the root cause of our dependency. We need to get to grips with what plastic is and why it's everywhere. This book distils the latest research, along with what I've learned about the plastic crisis over a career in writing, researching and broadcasting on the environment. I've interviewed industry leaders and environmental campaigners, talked to people like you and me who want to play their part and reduce their plastic consumption. Not only is this strangely fascinating (I warn you, it doesn't take much to become a polymer nerd, holding containers up to the light in order to try and determine their chemical content), but it is crucial in order to see the full picture and understand what is truly at stake. By exploring and increasing our understanding of the backstory of our plastic dilemma, we will become more resilient, resourceful and, I hope, more resistant to the avoidable plastic that blights our lives.

I take a look behind the curtain to explain what is really going on in our recycling system, and how and why the plastic that we generate is ending up as a pollutant. I uncover the way plastic is pushed upon us by retailers and manufacturers – behind everything we buy is a complex supply chain, and while we don't need to be expert on the entire length of that chain, as consumers we need to recognise the pinch points, where and why plastic is being added to the equation – and how we can take action to put a brake on it.

Part 2, New Tools, New Rules, is geared to helping you reduce your plastic footprint. I describe some of my wins, and outline down-to-earth ideas and strategies to devise your own Plastic Survival Plan. Here you will find a wealth of sound, simple tips and practical how-tos that you can put to work to make a change. Right now.

My plan is to get right between you and your plastic dependency and consciously uncouple your life from the material. In effect this is an enforced break-up, an intervention. We are so dependent on plastics in our lives, and we use so many habitually, it's incredibly rewarding to see how making small changes can yield big results. Every step is geared towards making sure we turn that tide.

The small steps I outline here might feel like a drop in the ocean but together, I believe, we can and will effect a change. It feels good to know you have an army behind you, doesn't it?

PART I

1

MY PLASTIC LIFE

Imagine that. A planet covered in plastic. We're already accustomed to seeing beautiful views ruined by plastic litter. But, as horrible as it is to imagine a carpet of fizzy drinks bottles and crisp packets covering land and sea, this is not merely aesthetic. The threat posed by plastic that has ended up in the ocean has recently been the subject of extensive media attention and growing public concern: of course, millions of tonnes of plastic are also polluting the land and soil. But it is the sea where scientists and campaigners are focusing many of their most urgent efforts: not only is it very hard to get rid of plastic once it is in the marine environment, but, as research emerges, it seems there are some serious threats that must be tackled now.

'Everywhere we have looked we have now found plastic,' Professor Richard Thompson had told me on my visit to his Marine Litter Research Unit laboratory at Plymouth University. It's a statement that continues to haunt me.

Microplastics have been found on beaches from Fernando de Noronha in the mid-Atlantic to Antarctica. They've even been found in the Mariana Trench, the deepest part of the global oceans.

Over the last twenty years we have truly been plumbing new depths, in the form of submersibles and ROVs, remotely operated vehicles that crawl along the deepest trenches of the ocean and record what's going on in this incredible universe 6,000 m below. This has changed our thinking. We now know that the Abyssal and Hadal zones of the world's oceans are not the cold, dead zones that they were once imagined to be, but biodiverse ecosystems, home to coral and hosts of living organisms. At this depth, life is slow: it takes a lot of time for plants and animals to grow and replenish. Contaminants in the form of plastic pollution cry disaster for this delicate ecosystem.

There is still research to be done to get a real understanding of the impact, but if plastic fragments are now everywhere, and culturally the material is so embedded in our patterns of behaviour and our lives, then you might think my suggestion that we can turn the tide is somewhat optimistic. You might also ask what business do I have, trying to recruit you to the cause?

Over the last few years I've been on what can only be described as an epic plastic adventure. I wanted to discover the extent of plastic's grip on our home lives, but also on commerce and in culture, and to understand its environmental impact and the implications for future generations.

I've interviewed plastics apologists, deniers, enthusiasts and lovers. Among them plastic is often referred to as the 'skin of commerce' – the implication being that whatever the downsides, we can't get by without it. I've always thought their assumptions worthy of interrogation.

Don't get me wrong: I was as immersed in the Plastic Age as anybody else. In fact, we're all children of the Synthetic Century. In common with most other kids born in the 1970s, I spent my childhood obsessively building with those small, coloured blocks of acrylonitrile-butadiene-styrene (ABS) more commonly known as Lego. Yes, I was the archetypal Lego brat, screaming blue murder if any of the monolithic structures I had carefully built were dismantled or damaged in any way by Hoovering. Only my beloved Barbie surpassed my love of Lego. Barbie's general sophistication and glamour, to my seven-year-old eyes, eclipsed the fact that she was essentially a number of different bits of plastic. Or, to be more precise, a rotationally moulded co-polymer for the arms; a body of ABS (like my Lego bricks) and hair made of vinylidene chloride polymers (known in the trade as Saran).[8]

Now, things are changing. Kids will soon be able to add garden foliage to their Lego landscapes made entirely from plant-based plastic sourced from sugar cane.

But, probably like yours, my childhood was relentlessly plasticised. When my neighbours brought me back a Mickey Mouse cup from Disneyland Florida with a twirly plastic straw, I thought it was the best thing since sliced

8. Craftech: 'Which plastic materials are used in Barbie dolls?' http://www.craftechind.com/plastic-materials-usedin-barbie-dolls/

bread – which I was also very fond of. Imagine my surprise, then, when I went to stay with my grandparents for the summer holidays, to find that my grandad did not share my enthusiasm. He shook his head. 'It's terrible that we waste plastic on stuff like this,' he said.

This was an unusual attitude circa 1981, and I remember it well, not least because it was a rebuke from grandparents from whom I usually had a 100 per cent approval rating. I noticed my grandad also took a dim view of the plastic laundry liquid dispensers then coming on to the market and relentlessly advertised on TV. In fact, he seemed to take a dim view of every brilliant consumer product lavishly constructed from plastic. Grandad was vocal on the subject, and told me in no uncertain terms that plastic was made from oil and that once you made something from plastic, it would take hundreds of years to degrade. It was clear that he thought this was an enormously bad idea.

Not only did he dislike the material, he seemed to make huge efforts to stop it getting into his bungalow, which, alarmingly, he declared a plastic-free zone. When we walked into town, he would loudly decry the use of polythene grocery bags at the checkout, loading his shopping into string bags that he carried everywhere. To my utter mortification, on some trips he actually unpeeled the plastic wrapping from his grocery shop and *left it at the checkout*, an act of rebellion covered by the *Chester Chronicle* on one memorable outing.

I know you're thinking, *typical hippy type*, but that wasn't exactly it. Before he retired, my grandad worked as a scientist for Shell, the global oil company. He was hardly J. R. Ewing, but even so, given that plastic is made from oil, and inextricably linked to the fossil fuel industry, his professional career meant that he was heavily invested in oil, so his stance was highly unusual. It also had an effect on me. My grandad's total and public rejection of plastic at the supermarket till may have filled me with horror at the time, but a seed had been planted. It took a while to germinate into a real interest, as I motored through my teens and twenties consuming fast fashion, fast food and generally living it up in a whirl of consumerism. But now, all these years later, I have to concede that he was on to something.

THE ECO AGONY AUNT

I'm fascinated by the flow of consumer goods into our lives and how that changes the earth's prospects. That means for many years I've been carefully scrutinising this rapid transit of 'stuff', then trying to figure out where it comes from and what the impact of this 'stuff' is. Somehow my strange pastime has transitioned from a hobby into my profession. As a journalist, in essence my job is to find out the true cost of stuff. So it's no coincidence that in 2015 I helped make a documentary, *The True Cost*, finding out the real cost of a collective addiction to fast fashion and bulging wardrobes. Spoiler: the results aren't too pretty.

With a weekly column in the *Observer* magazine, in 2004 my job description morphed into what can only be described as an agony aunt for the planet. If I had questions, I found that many other people did too. Running for thirteen years, we swept across the most extraordinary range of topics, from huge global issues such as 'Will corruption charges in the Brazilian political class affect the efforts to preserve the rainforest?' to the more prosaic, 'Can I put a margarine tub in my recycling?' To be fair, the recycling question was more my kind of investigation. But to my surprise, no thanks to the extraordinary array of local authority recycling schemes across England, Scotland, Wales and Northern Ireland (over three hundred at the last count – all subtly different and with some variations within the same postcode!), it was often more difficult to come up with a cohesive answer on recycling plastic containers than on the biophysical effects of corruption on rainforest biomes thousands of miles away.

In the early noughties, activism on climate change was in vogue, magazines ran special issues and celebrities seemed increasingly keen to become the public face of environmental campaigns. This was not universally applauded by the green community, which was pretty puritanical and had a major problem with the hypocrisy of actors and singers flying about the world speaking on climate issues. I get the point, but in Western society there are very few non-hypocrites, and I happen to find it really commendable when A-listers fast-track and amplify environmental messages. Let's face it, they do have an appeal that your average eco warrior doesn't.

Besides, on many issues I found the stars to be deeply committed. On one memorable occasion, actor Woody Harrelson came into the *Guardian* offices for a meeting about our annual Ethical Awards. After a bathroom break, he returned to the boardroom, furious, brandishing the soap dispenser, which evidently he'd just wrenched off the wall. To Harrelson the antimicrobial sanitiser from a detested multinational was beyond the pale, on the grounds that it was disruptive to the ecological system and bad for human health. We meekly promised to make some changes.

HOW I BECAME A PLASTIC DETECTIVE

From 2004, the ten years I spent answering my readers' questions was a hugely exciting and galvanising period for a new form of sustainability that was bringing together thirty years of earth science and mixing that data with ethical consumerism. We were really trying to develop a new blueprint – or 'greenprint' – for life, one where every individual could make better decisions; and place the planet at the centre of those. At heart, sustainability[9] is pretty simple: it's about leaving the earth in the same or better shape for the next generation.

9. But sustainability is also open to interpretation, and suffers from a lack of legal definition and precision. Sustainability can also be prone to over-claiming, particularly by global brands whose advertising campaigns would lead you to do some thinking. You may hear them claiming that they have single-handedly discovered a way to stop global warming, when they're actually just bringing a new chocolate bar or facial scrub to market.

The territory of over-claim and obfuscation is called greenwash – a term coined by Jay Westerveld, a journalist who, in an essay, noted the hypocrisy and irony of notices put up by multinational hotel chains informing humanity that by not washing a towel you are Mother Teresa and Gandhi rolled into one. You need to be on your guard, especially when it comes to turning the tide on plastics, because there is a good deal of 'plasticwash' going on.

And yet, no one seemed to be listening, particularly when it came to plastic. Almost all plastic is made from oil. Let's be in no doubt: plastic is a fossil fuel product, and it needs to be considered in this context.

BIG OIL

I have a habit of referring to different types of plastic as 'oily'. Some, like slimy cling film and rigid plastic boxes, feel more oily than others. But really all plastics are oily. In fact, of all the plastic we consume, 90 per cent is virgin plastic, made of oil.[10] In effect, this is a double insult to the environment: not only are we choosing to make new plastic instead of recycling what we've discarded, but to compound matters, we are also making it from fossil fuel.

Making plastics requires the heat-cracking of fossil fuel feedstocks, usually derived from crude oil, which is then converted into reactive hydrocarbons. After that comes polymerisation. Chemical additives are a key ingredient, giving plastic its different characteristics so that we can tailor it to our heart's desire.

At the current rate of production, for every barrel of oil extracted from the earth, 8 per cent becomes plastic: 4 per cent is the raw material used to make the plastic and the other 4 per cent is the fuel that powers the polymerisation

10. Ellen MacArthur Foundation, Project MainStream, World Economic Forum, McKinsey & Company: *The New Plastics Economy, Rethinking the Future of Plastics*, 19 January 2016.

process.[11] Plastic binds us to the fossil fuel economy, linking us directly to resource-conflicts and climate change. Our mission should be to decouple from oil with urgency. So while that 8 per cent figure might sound insignificant, isn't it counter-intuitive, at the very least, to be moving in entirely the wrong direction?

In an ideal world, of course, everybody would be working together to make sure that we curb our dependency on fossil fuels. But on planet plastic, the energy is moving in the opposite direction. Together the fossil fuel, plastic and chemical industries – hand in glove – are gearing up to unleash even more plastic on the world.

GET READY FOR THE GLUT

In the next five years, America is set to surpass Russia to become the world's biggest oil producer. You'll have heard of fracking, the shorthand for hydraulic fracturing. This is the process in which rock is fractured by a liquid injected into it under high pressure in order to force open existing fissures to extract oil or shale gas. In parts of the UK, such as Sussex and Lancashire, plans to frack have been met by community resistance and blockades. In the US, large-scale fracking has led to a shale gas boom, and the price of conventional oil has plummeted as shale gas floods the market.

You might be wondering how on earth a fracking boom

11. Hopewell, J; Dvorak, R, Kosior, E.: 'Plastics recycling: challenges and opportunities', *Philos Trans R Soc Lond B Biol Sci.* 27 Jul 2009; 364(1526): 2115–2126, doi: 10.1098/rstb.2008.0311

in the US has any connection to plastics in our lives in the UK. Well, it does. The shale gas boom is almost certainly about to be converted into plastic production that will flood the world market with new low-price plastic products, at a scale and volume never seen before.

The shale gas flowing in from West Texas is used as feedstock for ethylene, the building block for most plastics. And the US, already awash with more plastic than it knows what to do with, is aggressively looking for overseas markets to fill with this oncoming deluge of cheap-as-chips plastic. As chemical companies making plastic resins along the Gulf Coast scramble to find new markets, we consumers need to be armed and alert, and ready for the onslaught.

Fossil fuel industries have already taken a hit with the growth of electric vehicles (EVs). As transport severs its connection with petrol (something that seemed unthinkable just a few short years ago), the oil industry is looking for another outlet. Propelled by shale gas, polyethylene production most commonly used in packaging – where we already have a big problem – is about to take off.

Because there is no doubt: since 2010, $180 billion has gone into new plastic manufacturing plants across the Atlantic, and that translates into an almighty push to sell billions of pounds of extra polyethylene.[12] The world's most common plastic, used to make bottles and containers, as well as other common-or-garden plastic products, polyethylene is the plastic that already features most heavily in our lives. And in

12. Taylor, M.: '$180bn investment in plastic factories feeds global packaging binge', *The Guardian*, 26 December 2017.

our rubbish bins, and in litter lining streets and motorway verges, and in the ocean and on the world's beaches.

Make no mistake, we have a fight on our hands. But don't be intimidated: you are about to become the frontline of resistance.

Even if I'd wanted to, I couldn't have ignored the creep of plastic. Increasingly the ubiquity of plastic packaging and the inability or unwillingness of retailers to tackle the influx was driving my readers crazy. More than half of my *Observer* postbag was taken up with recycling and packaging conundrums. On a literal level, it was also hard to ignore: every Sunday the magazine containing my environmental musings would be generously wrapped in a particularly annoying plastic film. This did not escape the attention of readers, or indeed *Private Eye*, who mentioned it frequently. In fact, I've never been able to get a breakthrough on this. It is the supermarkets that have driven the wrapping of Sunday newspapers in plastic to stop all the hundreds of advertising inserts – a pet hate of many readers – falling out.

The old journalism adage, 'follow the money' became, for me, 'follow the oil', and as we now know, a lot of that oil becomes plastic. So I followed the plastic, sometimes literally. I pestered CEOs in boardrooms and I staked out landfill dumps; I joined beach cleans and I helped to release turtles back into the wild. Some of this I've documented in newspapers and on TV. All the time my readers kept writing to me with a certain amount of anger about the surfeit of plastic in society. I heard from many women fed up with plastic bags – given out liberally in all UK supermarkets –

attaching themselves around the heel of their shoe on a windy day walking down a high street.

From time to time I would get to debate plastic pollution with representatives of the plastic industry's members' organisation, called the British Plastics Federation and the Industry Council for Plastic and the Environment, INCPEN. My adversaries seemed to me to be perpetually bad-tempered, or perhaps it was just that they certainly made it clear that I was an irritant (far more annoying than plastic pollution). In one surreal exchange I took part in on BBC Radio 5, a plastic industry representative tried to argue that a plastic bag levy would be downright dangerous and actually increase the amount of plastic that society used. The plastic industry allied with major retailers and manufacturers who were heavy plastic consumers. The message was loud and clear: they were producing plastic packaging in all its glorious formats, from plastic bags to takeaway cartons, because the consumer wanted it. Plastic packaging was convenient for all for us, and ungrateful naysayers like me needed to get back in their box.

I realised that I had to prove that most consumers did not want excess plastic in their lives, and that many were also extremely angry about having to deal with it. Back in December 2005, I persuaded four families to save their rubbish for a month to demonstrate exactly what they were throwing away over Christmas.[13] Every piece of waste the families collected was weighed and analysed, and plastic

13. Siegle, L.: 'One family, one month, 50 kg of packaging. Why?', *Observer* Magazine, 29 January 2006.

was by far and away the dominant material. We were particularly alert to over-packaging, where plastic had been used unnecessarily.

One of the most horrible examples we found was a shrink-wrapped coconut from the retailer Morrisons. Given that coconuts famously arrive in their own protective shell, I argued this was unnecessary waste. The retailer fought back on two points: first, that the plastic film was necessary in order to attach a metallised sticker with a barcode, and then – when this didn't fly – that the fibrous hair of coconuts might be inhaled by customers, and therefore constituted a health and safety hazard. We talked. Eventually Morrisons agreed that they would stop shrink-wrapping their coconuts.

Any sense of victory was short-lived, however. The following week another reader contacted us to say that the retailer had moved on to shrink-wrapping cucumbers. I began to feel like a crusading greengrocer, forever doing battle on unnecessarily plastic-packaged fruit and vegetables. But I did learn from these encounters. I realised then, in a way I hadn't before, just what a grip plastic had on our everyday lives, and I started collecting data that would eventually lead to strategies to help households stop the flow into their lives. It is these experiences that have set me on the road to the tips and advice laid out in the pages of this book.

One of the things that emerged for me from working with the families back in 2005, and every other time a household has been kind enough to let me move in and root through their dustbin, is just how complex our recycling system is. While we all have a moan about different boxes and bins and a lack of harmonised recycling in the UK, it seemed obvious to me that some of the actual plastic wrappings on everyday products had become so complicated that you needed a PhD in polymeric science to understand which bin they should go into (for more on recycling, see chapter 14, on page 199).

I came across a piece written by and for the plastic packaging industry that shed some light. 'The average consumer probably has no idea that the packaging of a typical product he or she might pick up weekly may have as many as six layers of plastic (even more are quite possible) and can sit on the shelf and remain fresh for several months, possibly up to a year.'[14] The penny dropped. It was not for our convenience – no consumer in their right mind wants food in their refrigerator for a year – but for the back-of-store convenience of retailers. Plastic packaging was being applied with increasing zeal because it was cost-effective and made life easier for retailers and manufacturers, who neatly argued that it was ultimately in our interest because the packaging resulted in lower food prices overall.

This may well be true in part, but we were, nevertheless, left to deal with the consequences: 90 per cent of the cost of collection, sorting and disposal of plastic packaging is borne by us, the householder, through our taxes. Meanwhile the burden on the environment is incalculable. Key to turning the tide of this waste, and one of our missions through writing this book, is to level that playing field.

Our plastic problem had become entrenched, and extremely

14. Koltzenburg, T.: *High-Barrier Packaging: Yesterday, Today, and Tomorrow*, 1 February 2000.

complicated. Unwrapping and shedding the plastic meant feeding into a complex, patchy waste infrastructure with a multitude of recycling systems and bins and boxes which confused almost every householder I met.

But believe it or not, there was also a lot of effort being expended on solving some of these problems. In the UK the government set up special advisory panels and organised innumerable conferences on minimising waste and designing better plastic consumables. There was a huge amount of energy from 2005 to 2009 as part of a voluntary initiative called the Courtauld Commitment. This was a series of agreements between leading grocery retailers and manufacturers, WRAP (the Waste Resources Action Plan) and the four UK administrations – including the Scottish Government. Now when I look back these were halcyon days. The focus was on new solutions and technologies that would reduce the amount of packaging of all materials getting into our bins – but of course the focus became plastic. This was a voluntary agreement, but nevertheless the results were impressive. According to the official statistics, over this four-year period 1.2 million tonnes of food and packaging waste was prevented, a saving of 3.3 million tonnes of CO_2 emissions, which is equivalent to the emissions from 500,000 round-the-world flights.

WRAP was launched in 2000, and has continued to play an important role in pushing our country from the bottom of the international recycling tables (where it once languished) upwards. In 1996, we were at the bottom of the European recycling league table, alongside Ireland and Greece. We recycled a minuscule 7.5 per cent of our rubbish. But the good thing about being at the bottom is that the only way is up, and that's the way we went. Statutory targets were set in 2000, obliging councils to offer recycling. Then the Household Waste Recycling Act became law in 2003, requiring every household to be provided with a doorstep recycling collection for at least two materials by 2010. In 2005 there was even a £10 million national TV advertising campaign to make us get into the habit of recycling.

By 2007, we were each recycling 171 kg of rubbish a year, securing mid-level status in the league table.[15] Then, wouldn't you know it, the wheels started to come off. By 2008, when the global recession impacted, I reported that the bottom had dropped out of the recyclable materials market. Stockpiles became commonplace as councils struggled to shift materials, and some had to be housed on Ministry of Defence bases until manufacturing picked up. Eventually it did, and the threat receded.

WRAP also brought the major food retailers who dominate our grocery landscape to the table, and they promised to make lowering plastic consumption a big priority. I had confidence that they'd make it easier for us to do a food shop one day that was pretty much unpackaged.

I might have saved my energy. If I have a criticism of this period, it's that I gained a lot of insight and then gave up too easily. Those of us who were pushing for better recycling, smarter design of everyday items and less single-use plastic

15. Shreeve, J. L.: 'Recycling industry in a slump', *The Telegraph*, 14 November 2008.

– and there were many of us – ceded too much control to the retail, manufacturing and plastic industries. They in turn came up with a typically complex system of levies that meant the biggest producers of the plastics that enter our bins – the big manufacturers and household brands churning out billions of bottles, tubs and trays a year – do not pay the true cost of production, impact, collection and recycling of those products. Nor do they pay a penalty if the plastic they inflict on the world is not recycled.

HEARTS AND MINDS

I began to realise that for many people, plastic pollution is an extremely emotional issue. It affects them in a way that climate change doesn't always (the challenge of communicating the danger of an atmospheric gas that you can't smell and aren't aware of was always going to be a tough call). I was moved to tears watching a Sky News special. A man in his sixties showed a reporter the plastic that had washed up on a beach in the West of Scotland. He had lived there all his life, and matter-of-factly pointed out the typical dystopian hillock of water bottles, tampon applicators and crisp bags tangled up with seaweed on the shoreline. He pronounced it 'a disgrace'. Then he began to cry and I found myself crying, too.

We weren't crying for ourselves, but for our grandkids, or someone else's grandkids. I don't have children myself, but as an environmentalist I get the opportunity to fight for yours. Ultimately we're all headed in the same direction. We're all trying to avoid crushing, abject failure. And, if you're looking for the definition of human failure, it is surely bequeathing to future generations a planet trashed beyond repair. It would add insult to injury if that should occur through making bad choices about stuff like plastic food wrappings. It doesn't bear thinking about.

How would we suffer the indignity of explaining to others that we allowed ourselves to be convinced that there was no harm in making single-use items like ketchup sachets and drinking straws out of plastic, a material that would last and pollute for centuries?

2

THE BIRTH OF THE AGE OF PLASTIC

One of the saddest concepts of the green movement is that of the 'shifting baseline'. Each generation has a mental image of a baseline of how the world looked in their youth. In our mind, we compare any change in our environment and in our life to this baseline. Younger generations accept as normal a world that, to older generations, seems tainted and degraded. Transposed to the plastic pandemic, today's level of plastic waste pollution on and off land will soon – and probably already does – appear normal to kids and young adults. Images of turtles tangled up in the loops of plastic that hold a six-pack of beer together, footage of a whale ingesting plastic or of a diver in Bali swimming in the plastic soup will not only lose their power to shock, they will quickly appear normal.

The man who coined the phrase 'shifting baseline', Callum Roberts, Professor of Marine Conservation at the University of York, is in himself very jolly, so I can say with some certainty that he didn't come up with it to depress us all and make us feel hopeless. Rather he wanted to articulate urgently that we must stay alert to this phenomenon. The 'societal amnesia' of the shifting baseline lowers our expectations and our level of ambition when it comes to protecting and restoring nature. There must, however, be no societal amnesia about the scourge of plastics. We must bang the drum loudly and never stop.

It's clear: if you love wildlife and nature, you need to kick as much plastic out of your life as possible. So it's a tremendous shock and a considerable irony that the initial appeal of plastic, and one of the reasons why it was invented in the first place, was to take the pressure *off* wildlife.

BEGINNINGS

I recently made a pilgrimage to Hackney in east London, where I stared up at an unassuming brick wall. My gaze alighted on a plaque, in the familiar corporation font, arching around the Hackney civic crest. It read:

> *FIRST PLASTIC IN THE WORLD. Known as 'Parkesine', invented by Alexander Parkes. First made near this site, 1866, at the Parkesine works.*

Being a plastics nerd, I would take issue with the London Borough of Hackney because in actual fact Alexander Parkes first filed his patent for Parkesine in 1855, when he was still living in his native Birmingham. But let's not split hairs, or indeed polymers. Alexander Parkes was a prolific Victorian inventor who no doubt would have been bemused by the very fact that there is a plaque commemorating Parkesine, just one of his many innovations, in Hackney. Out of the sixty-six metallurgic patents and fourteen patents in other materials, such as rubber and waterproofing formulations that Parkes filed, Parkesine was in many ways the least successful during his lifetime. The waterproofing process that he sold to Mackintosh, the famous raincoat brand, earned him substantially more plaudits, and money, while he lived. So, he may have thought, why a plaque for this one?

THE ANIMAL'S FRIEND

Although Parkesine wasn't Alexander Parkes' most successful invention, the opening of the Hackney factory addressed an emerging need. Today, the London E9 postcode has been colonised by the hipster community, but in the late 1860s it would have been a thriving hub of garment manufacturing and button production. The garments of the era were complex constructions: shirts required separate, stiff collars and everything required a huge number of buttons, which were traditionally fashioned from cow horn, mother of pearl, tortoiseshell or ivory. This put a huge strain on natural materials as, all across the British Empire, wild animals were hunted down and killed so that great quantities of buttons could be carved in Hackney Wick.

No tortoises, however, were in fact harmed in the making of buttons, rather the poor hawksbill sea turtle: 'No shell has been put to greater uses than the tortoise-shell, which has nothing to do with the tortoise,' the *Lancaster Gazette* advised readers in 1885. 'For the tortoise-shell of commerce is derived from the beautiful horny plates of the hawk's-bill or imbricated turtle (*Eretmochelys imbricata*), though from those animals only that weigh at least 160 lb as the plates are otherwise too thin'.

The Victorian appetite for tortoiseshell was insatiable; hair combs, belt and shoe buckles, snuff and trinket boxes, picture frames and jewellery were all made from the material. Not since Ancient Rome, when the hawksbill shell was in demand as a baby bath, had the population been under so much pressure. By targeting the bigger, fully grown turtles, which were harpooned like whales, hunters were removing the kingpins of their population. The fact that the English also tended to see the species as floating bowls of soup, added to the issue. In the 1880s the London *Evening Standard* bemoaned the extortionate price of turtle soup due to a scarcity in the wild and cheered on attempts to establish large-scale hatcheries. Today, unsurprisingly, the beautiful hawksbill turtle is critically endangered.

For a few years, until the factory closed, Parkes found a market for his Parkesine as a man-made substitute in collars, cuffs and buttons. His plastic invention offered the wild hawksbill population a reprieve. Without Parkesine – and the subsequent plastics by which it was soon eclipsed – would the hawksbill sea turtle have become extinct?

THE PLASTICS RACE

The early plastic pioneers were the archetypal geniuses in sheds, their methods conforming to Edison's idea of achieving scientific greatness and transforming the world through 'one per cent inspiration, 99 per cent perspiration'. Sometimes the journey to discovery was just plain bonkers, laden with risk in a way that our modern safety-conscious society doesn't really do. Contemporaneous reports describe Parkes preparing inflammable solvents and dissolving them in fire-raising phosphorus in a basement 'laboratory' where he somehow avoided 'serious accident'.

The race to make plastic on a manufacturing scale was intensely competitive, giving the whole process a dynamic story arc with characters that we possibly learned about in school science lessons, such as Sir James Swinburne, who took plastic to the next stage and made it fireproof, and later Leo Baekeland of Bakelite fame, who is also frequently referred to as the 'father of plastics' and who created the first thermosetting plastics that could be heated and shaped.

The very first wave of plastic products must have seemed like modern miracles to those who saw them. As with Parkesine and tortoiseshell, the earliest uses of the new invention were as a substitute for more traditional materials – many of which seem to us today downright repulsive.

As the nineteenth century progressed, an expanding American population also created a demand for natural materials. In the 1860s a billiards craze was sweeping America. It required fifty tusks from the Asian elephant in order to find one with the requisite smooth ivory, and just three or four billiard balls could be carved from each tusk. The early adopters of conservation worried that there would be a wild elephant shortage. In fact, this was realised, but some decades later. Ivory also became prohibitively expensive. Enter two stars of the sport, Michael Phelan and Hugh Collender, Irish-born American entrepreneurs who were both players and owners of billiard parlours. They evidently had a gift for marketing, too. In 1863 they placed an advertisement, setting down the challenge to invent a substitute for ivory and promising $10,000 in gold to the successful inventor scientist.

While there is no firm evidence that Phelan and Collender actually paid out, by popular assent a US inventor named John Wesley Hyatt was the winner of the challenge. He may not have claimed the billiard-ball gold but in 1869 his work on synthetic celluloid simplified the production process and brought us one step closer to 'modern' plastic. His invention also offered an alternative to the killing of thousands of wild elephants for their tusks. I think we can be pretty grateful for that. Hyatt's material, patented as Celluloid, later found an application for use as dental plates. When used for billiard balls, the results weren't always predictable. Nitrocellulose, also known as guncotton, was highly combustible. The billiard balls had a tendency to explode.[16]

16. Bijker, W.: *Of Bicycles, Bakelites, and Bulbs: Toward a Theory of Sociotechnical Change*, MIT Press, July 1995.

Taking a historical perspective on plastics also allows us to acknowledge the groundbreaking nature of the innovation in the race to produce and commercialise plastics. Backstreet-genius chemists like Alexander Parkes and John Hyatt worked up solutions in basements and lean-tos that would have given health and safety inspectors of today major cause for concern. But make no mistake: this was breakthrough chemistry. It was subversive and extraordinary, as they moved away from the confines of classic organic chemistry. They were in effect making something that was entirely synthesised by their own hand.

These early experiments in plastic showed that humankind didn't have to be dictated to by nature. Limits weren't set by using wood from trees or ore dug up from the ground where the behaviour, amount and structure of the material was already dictated. Instead chemists were able to tamper and alter the molecular chain of plastics, giving the material different properties. It could bend, stretch or become translucent or incredibly durable. It put the chemists in control.

As the chemistry progressed, substances that originated in a chemist's test tube, rather than mined from the earth, were shown to be stronger than steel. Plastic could be produced with specific properties, such as extreme heat resistance. By the 1920s seemingly eccentric, individual discoveries – celluloid, cellulose, acetate, phenolic, amino plastic and nylon – had started to be grouped together as the family we now call 'plastics'.

At this point, the production and process of polymerising

molecules – the science of making big molecules out of small ones – must have seemed incredibly exciting. Here, for example, is a description of a plastic ruler from the *Illustrated London News* in 1943:

> *One of the early examples figured at an exhibition of modern glass-making, where a stick of it as long as an office ruler could be looked through all its length, as if it were without substance.*[17]

It also had a noble purpose. This is from the same correspondent, a description of the uses of plastic that must have had particular resonance in 1943 for a nation at war:

> *Other examples of transparent plastics are the moulded shields that protect the bomber pilot or the domed cupolas of the powerful craft which speed to pick him up if he bails out into the sea.*[18]

The use of plastic wasn't just clever or chemically innovative, it was saving lives and helping to secure an Allied victory for the Second World War. If that's not good PR that is likely to have long-term appeal and settle into a nation's psyche, then I'm not sure what is. By the 1950s the commercialisation of plastic had arrived, and the stage was set for an incredible plastic binge that was to sweep up several generations of 'consumers'. You could argue it was the start of the plastic pandemic. Photographs from 1950 show two scientists, Karl Ziegler and Giulio Natta, proudly holding small bins and buckets made of Hifax – their version of plastic. In retrospect, the products they're holding in the pictures

17. Grew, E. S.: 'The Plastic Future', *Illustrated London News*, 26 June 1943.
18. Grew, 'The Plastic Future'.

might not look particularly world-changing or exciting, but nevertheless Ziegler and Natta bagged the 1950 Nobel Prize for their work on polymerisation, the process of making plastic. You can't get more 'legit' than that.

THE RISE AND RISE OF PLASTIC

Once legitimised, the commercialisation of plastic took off, and the material became allied to a post-war consumer boom. Once we took to plastic, we fell under its sway with indecent haste. In just a few short years, a make-do-and-mend domestic culture passed down from one generation to the next had been turned on its head.

The polyethylene bag, unknown to pre-war generations, is a symbol of this decisive shift. By 1960 over twenty million plastic bags were being produced in the UK. They went from zero to ubiquity in the space of ten years.

Then there was plastic packaging for consumer goods. By 1970, the UK was ploughing through 350,000 tons of plastic packaging every year. The last threads that linked us to grocers in brown aprons, who wrapped purchases in brown paper, were severed. The thin blown plastic film we now know as Cling Film, or Saran wrap in the US, became a substitute for greaseproof paper. Vegetables were pre-prepared and packed into plastic trays as society began to eat as if it was permanently on board a Pan Am flight. No longer would you select and pick nails and screws; they would be ready apportioned and captured behind a moulded pack.

Then there was the posh plastic packaging that emerged (for me, this really conjures up Mike Leigh's seminal play *Abigail's Party*) conveying, improbably, some sense of luxe to whatever was encased inside. In order to be really posh – say, for a luxury product like bath crystals – the plastic jar could even be fluted, so it resembled glass.

If consumers loved it, the branding and marketing people loved it even more. When consumers could see something in a bubble or tamper-free pack, they were more inclined to buy it. It is this drive and enthusiasm for all things plastic that has got us to where we are now. Which is where, exactly?

3

THE WAKE-UP CALL

Over the course of my career, I've had the opportunity to do some incredibly cool things. I've seen atolls teeming with whale sharks, sailed on the iconic Greenpeace vessel *Rainbow Warrior*, and lived with an indigenous community in the far reaches of the Brazilian rainforest biome on the border with Venezuela. But the only time my friends have truly seethed with jealousy is when I met Sir David Attenborough.

For our first meeting I rang Sir Attenborough's 'office' to arrange an interview, and to my immense surprise he answered the phone, his inimitable authoritative whisper catching me completely off guard. 'Oh, yes,' he said, 'would next Tuesday suit?' I was, of course, reduced to a gibbering wreck. On that occasion, the thing I remember most was his talking me through the fossil collection housed in his living room. He's been collecting fossils since he was a young boy, and that sense of wonder was still very much intact.

Attenborough's seminal BBC series *The Living Planet* was

broadcast when I was at primary school. To people of my vintage, it was like a form of magic. Glued to our screens, we sat transfixed, with Attenborough as our guide, in awe of the world's biggest trees and learning about the rain cycle, or witnessing birds with incredible plumage and meeting indigenous communities who lived in nature. It certainly didn't mirror the suburbia I was used to, but that voice somehow made it very immediate. I began to understand that all of it was interconnected and functioned together. The 'living planet' was an ecosystem, and it was my home.

In October 2017 an audience of fourteen million watched the BBC series *Blue Planet II*. That huge audience was confronted by the reality of plastic's grip on us, and our addiction to it. The famous underwater cinematography showed us a pilot whale with a bucket in its mouth, an albatross feeding plastic to its chicks and a dolphin potentially exposing its newborn calves to tiny pieces of plastic pollution. The potent images were accompanied by David Attenborough's powerful rallying cry to protect our oceans.

Something shifted. The stuff you might have seen out of the corner of your eye – the plastic bags caught in the branches of winter trees, the bottles and containers, flip-flops and fishing net tangled in the seaweed in the bay where you take your summer holiday – took on real meaning. Both our connection to this mess and our culpability for the impact on these sea creatures were laid bare. The plastic pandemic was brought to life. Our home was being trashed.

While we might balk at anybody else telling us what to

do, or suggesting that it's time we stepped up, when David said it the nation jumped to attention.

Making *Blue Planet II* was a mammoth operation. Crews were despatched on 129 expeditions to thirty-nine countries.[19] They did not set out to make a big deal of plastic pollution. In fact, as Executive Producer James Honeyborne explained in the aftermath in April 2018, the impact of the show, while thrilling, put them in uncharted territory. 'Our job is to tell stories about incredible marine life, and that's what we set out to do, to help people feel connected to life beneath the waves, but when the crews went out [. . .] they would find plastic everywhere they looked. So it became part of the story, we couldn't ignore it. If we were going to give a contemporary portrait of the world's oceans, we'd have to include plastic.'

He added, 'It's very strange for us [as natural history film-makers] to see the reaction the series has had at a political level because it's not that we went out to make a campaigning film. We're not campaigners. In fact, we're there to tell people how wonderful sea life is.'[20]

The ubiquity of plastic in the environment and our everyday connection to it means that we cannot ignore it. It is in your face. Plastic is taking over our planet. In the natural world plastic has now encroached on every biome. It doesn't so much interact with the environment as disrupt and disfigure it, and much of the damage could be permanent.

19. Interview with James Honeyborne, Executive Producer, *Blue Planet II*, as broadcast 16 April 2018. *The One Show*, BBC1, Plastics Special.
20. James Honeyborne interview, *The One Show*, 16 April 2018.

It's an issue that moves us quickly from awareness to activism. It lends itself to individual action, too.

Blue Planet II put the issue front and centre, and suddenly we were all talking about one thing: plastic pollution. After the shock and dismay came incomprehension. How was so much of our plastic waste ending up in the sea? After all, many of us had become dedicated and assiduous recyclers.

REAPPRAISAL

To me, it felt as if a starting pistol had been fired. There was a clamour for solutions and remedies. There was a great wave of energy as people tried to expunge plastics from their life: from the high street, from their cars, from their caravans – you name it. Hundreds of people got in touch with me on BBC Television's *The One Show*, on which I am a reporter and presenter. Following *that* episode of *Blue Planet II*, we were, I suppose, the natural focus for viewers' questions and suggestions, and they contacted us in waves. Before long our coverage of the topic evolved into a weekly segment on turning the plastic tide. Accepting that many plastics can't be avoided, and indeed are necessary, on the show we took aim at avoidable single-use plastics.

On one early show, Gordon Ramsay was our star guest. When I began haranguing him about banning single-use plastic straws in his restaurants, he viewed me with some amusement, before pronouncing that I was worse than his mother. It was a good-natured exchange, but I meant it. It was time for the big brands and the famous names to show leadership in our crusade against plastic waste. Some months later Gordon Ramsay Associates confirmed to the show that they were swapping to biodegradable straws.

It's always a win when one of us takes a stand by renouncing single-use plastics – but when it's a well-known celebrity with millions of young fans on social media, it's a double bonus. I was delighted, for example, to find that the singer and producer, Ellie Goulding, uses a metal beaker and paper straw on stage. Small changes, yes, but another victory nonetheless.

Through a 'green' friend, I got to meet Ellie, who is by her own admission 'on one' about single-use plastic and environmental concerns generally. She is a UN Goodwill Ambassador for the Environment, and in December 2017 we attended the third UN Environment Assembly in Nairobi, Kenya together. We visited a turtle sanctuary at the protected Watamu Beach on the Indian Ocean. It was an extraordinary experience as the marine biologist introduced us to the patients – two hawksbill turtles lucky enough to be picked up by fishermen who dropped them at the animal hospital. It was a typical story for the guys there: the turtle's guts get blocked with plastic and they are unable to feed, slowly dying. Death by trash. Once the larger of the two turtles was treated and the plastic successfully removed, I helped to release him back into the water, his flippers slapping against our legs as we went. He was strong, and back to full turtle health, but for how long? And what happens next time? As we released him into the surf, he passed a floating flip-flop as he propelled himself into the ocean. Incoming plastic; outgoing turtle. (Neither was the coincidence lost on me that this was the exact species granted a reprieve by Parkes' original invention of plastic back in Hackney.)

The next day Ellie, her manager Hannah and I took a stroll along the coastline. As we walked and talked, what started as casually and unthinkingly bending down to pick up a washed-up water bottle, turned into an epic beach clean with us staggering along the shore, laden down with bottles and snack packets and any number of bits of sandy plastic, including the tough plastic silver-grey envelopes that you'll know from online deliveries and a great number of old instant noodle packets. The origins were impossible to decipher; we could only guess at them. Some of it would have come from the shore, where there's a lack of infrastructure for rubbish disposal.

Back on the beach, our guest-house owner took the rubbish we arrived home with in good grace and promised to deal with it. They were sad but not surprised. Beach pollution is a fact of life here, on this beautiful natural coastline, and needs to be plucked from the long grasses that frame the sand. We passed several women out for their morning stroll in designer beachwear, dragging hessian sacks and filling them with plastic. The sack of rubbish has become this season's accessory by necessity. Certainly, we vowed that next time we visited, we'd come equipped.

REAPPRAISAL

I returned home to find that some changes were suddenly afoot. The outcry on social media and in the press in the wake of *Blue Planet II* had translated into something more meaningful. In early January 2018, the British Prime Minister Theresa May launched the government's 25 Year Environment Plan (the first time since 2003 that a British Prime Minister had laid out an environmental strategy in the House of Commons). 'A Green Future: Our 25 Year Plan to Improve the Environment' included a vow to eradicate all avoidable plastic waste over the next twenty-five years. *Blue Planet II* was cited specifically as one of the reasons for that pledge.

Days later, on 16 January, frozen-food retailer Iceland announced that it would eliminate plastic packaging for all its products – across 1,400 product lines – within five years to help end the scourge of plastic pollution. I was elated. The industry seemed less impressed, and the British Plastics Federation responded in the vein of Scrooge attending the office party: 'Growing and transporting food consumes a lot more energy than that used to make the packaging protecting it. Iceland's proposals target products that will have absolutely no impact on reducing marine litter, which in the UK typically comes from items littered outside our homes.'[21] Ouch, I thought.

21. BPF Response to Iceland's Announcement, http://www.bpf.co.uk/article/bpf-response-to-icelands-announcement-1261.aspx

Iceland was not at all deterred. When I visited their headquarters near Chester, MD Richard Walker confidently showed me the paper and pulp trays and paper bags he believes will take his brand's products to plastic-free status by 2023. In practice I found that Iceland's plastic-free packaging is already hitting the shelves. 'Not nearly enough plastic is currently captured by recycling in the UK, and until this year we were largely exporting the problem,' Richard Walker explained.

He is clear that he is in a position to shift his production to face off a crisis because Iceland remains a private, family-owned company. Other big retailers are answerable to shareholders, who are not apparently as stirred by the plastic pandemic. 'We're turning off the tap,' he said. 'It's the only way to get to grips with this.' A surfer with small kids, Walker is driven towards change. 'Otherwise what's the point?' he asks me. The noise that Iceland's Richard Walker has been making on plastic has only served to highlight the deafening silence of the other big retailers.

As energy and noise and announcements continued to swirl around the subject of plastic, how did this translate into real domestic family life? I wanted to see for myself, so I rang to ask the Proud family in South Manchester if they'd let me and a film crew go through their bins. They very kindly agreed.

The Proud family consists of mum Louise, who runs her own hairdressing business, dad Wayne, a design manager for a large building and architectural firm, their daughters Macey and Amber, and Ruby, a small, cute dog that can simultaneously growl and let you stroke her, and a tortoise. The Prouds built the extension on their house with their own fair hands. Now it resembles an upscale Ibiza interior, all white walls and opening out into a barbecue area, but in the depths of winter it was a story of mud and struggle. I note this by way of an insight into their character: the Proud family is not easily deterred. And yet when we set the challenge of reducing their plastic waste, it was tough going. They achieved a lot in a small amount of time, cutting their plastic use by 30 to 40 per cent in the first week. On the downside, they struggled to balance costs, and in seeking out plastic-free, their essential shopping became more expensive.

We made several films with the Prouds for *The One Show*. I loved having the opportunity to sit around their kitchen table and talk rubbish (!). It was not only a great insight, seeing the challenge through the Prouds' eyes, it was an essential piece in the jigsaw, showing how individual households can fight the tide of plastic.

The family have become avid plastic-reducers and were already avid recyclers, but when we followed the family's recycling to their local Materials Recovery Facility (MRF, pronounced 'murf' like 'Smurf') we found that a high proportion of recycling was not being turned into useful new items (as they'd assumed), but were being burned in energy recovery facilities. We were also able to take the family's concerns directly to the policymakers, the plastic manufacturers, the polymeric scientists and the campaigners and big recyclers, all of whom appeared on *The One Show*'s distinctive bright green sofas. It's not always easy to get joined-up answers on this topic, but we were determined to try.

After six months I was such a stalwart at the recycling centres, which included MRFs and even a PRF (Plastic Recovery Facility, pronounced 'purf') that in Skelmersdale they joked I could run it without them, and in Longley Lane in Manchester I was solemnly handed a business card with a drawing of a rat on it. This, I was reliably informed, was to be handed to my GP in the event that I became unwell, and explained some of the gruesome particulars of Leptospirosis, aka Weil's disease. It was a solemn reminder that rats are an occupational hazard of the waste business (I'm particularly phobic) and that this industry is gritty and real and not very glamorous, even for those of us who dip in and out for TV reports. But even so, we needed to get to grips with it.

THE DEEP DIVE ON RECYCLING

There's a persuasive line on recycling when it comes to the flow of plastic into our lives. Unsurprisingly, I hear it used by the plastic industry but also by retailers and politicians: 'Don't worry,' they chorus, 'we will just get better at recycling.' This silver bullet approach would be all well and good, if only it was that easy.

I love recycling, and I've always taken my own household recycling seriously. The process of altering the molecules in waste materials so that they can become something else seems magical to me. Energy is transferred from one matter to another in a continuous loop, and is therefore harnessed. That means that precious resources and the environment are conserved. When recycling works, it really works. Recycling plastic, for example, minimises energy and resource use by avoiding the extraction and processing of virgin oil, and it also reduces CO_2, particulate matter and other harmful gas emissions, compared to other methods of disposal or recovery. It also diverts our rubbish from landfill, and this is really important.

But in the UK, as we have seen, confusion still reigns when it comes to the question of what really happens to the waste that, each week, we dutifully collect and deposit in recycling crates and waste bins outside our home, ready for the bin man. It's a fascinating and complex story, but I won't go into the entire history of the UK waste industry here – I don't want to lose you just yet. I will, however, give you the highlights, as these shed some light on the how-what-where we are today vis-à-vis our rubbish.

As a former mining nation, disused tin, China clay and coal pits are dotted across the island. Historically these empty pits made an ideal place to pour in our rubbish. Landfill was convenient and cheap, so we literally buried our waste and the true cost of the issue with it. The fact that we could shovel our rubbish into gigantic holes meant that there was little incentive to find alternative means of waste management, and to some extent enabled us to turn a blind eye to ever more profligate levels of consumption.

Necessity is famously the mother of invention, and by

the end of the millennium, as we filled up more big holes, it became pretty clear that we were running out of space. It was also glaringly obvious that mixing hazardous, toxic materials with general rubbish in landfill had potentially lethal consequences, including noxious juices leaching into the environment. Waste entrepreneurs and experts were also cottoning on to the fact that much of the stuff being poured into this disgusting sludge was recyclable, and that by burying it in this way, we were losing its value.

In 1996 a new regulation was introduced to try to deter us from shovelling our waste into holes and move us towards recycling. This arrived in the form of a Landfill Tax – the original cost was £7 per tonne of general waste, with a lower rate of tax on 'inactive' waste such as stone, concrete and ceramics. Every year, the tax would be raised as we weaned ourselves off big holes in the ground and were encouraged to recycle.

Even if the will was there, however, the sustainable alternatives to landfill were not in place: there weren't enough recycling facilities, and there was a lack of information and education. We were dubbed the Dirty Man of Europe, a reputation which has taken a long time to shake off.[22]

Today, the landfill tax in England has risen to £88.95 per tonne of general waste, and against all the odds, we have got much better at recycling. When the figures for 2015 were released last year – it takes a while to aggregate them

22. Wheeler, K.: *The dirty man of Europe? Rubbish, recycling and consumption work in England*, CRESI WORKING PAPER NUMBER: 2013-01, Department of Sociology University of Essex, 'Consumption Work and Societal Divisions of Labour' project, funded by the European Research Council.

– the UK tied with Italy for tenth place, or mid-table in the European league, so not spectacular by any stretch, but no longer in Dirty Man territory.

As recycling has progressed, so has the technology. in the UK, we now have around a hundred MRFs, in the form of giant sheds into which flows all our mixed solid waste. Once inside the MRF, the priority is to sort repeatedly all the glass, metal, paper and plastic into what is useful and recyclable and what is not, as quickly and efficiently as possible. The MRFs are full of high-tech equipment, but human hands are still the traditional first port of call. When it comes to plastics, to the untrained eye the different resins can look very similar, but not to the pickers who snatch and separate the prized clear plastic bottles from the less-prized plastics – the cloudy bottles and flimsy yoghurt tubs – in what feels like nanoseconds. Even so, the number of actual people working in recycling is in decline, as the big waste operators make the switch to Artificial Intelligence (AI) robotic sorting.

How often do we stand at the bin, completely flummoxed by a piece of packaging? Can it be recycled or not? I know that in my house, it's a regular conundrum. Plastics are slippery. A piece of packaging might look like an easily recyclable foil, but in reality it's bonded with extruded polyethylene. It might look silky, like a fabric derived from a natural fibre – like teabags, which indeed were originally pieces of cotton muslin fabric – but in fact are made with polypropolene. At the MRF this confusion about what is recyclable is laid bare, and the plastics that are too hard and too costly to separate and process are also removed. Many of the containers and packaging that enter the MRF have very low recycling rates: according to WRAP, only 10 to 15 per cent of mixed plastics are recycled, including the rigid packaging used for supermarket meat and fish.[23]

Many MRFs have already invested in sophisticated sorting equipment. Ballistic separators bounce and vibrate waste along a belt, with the heavy plastics made from different resins falling down between grate bars onto another belt below, where they are channelled into another waste stream. During the process, pumped air separates lightweight plastics from the heavier glass and metals.

Once captured, many facilities employ near-infrared technology to sort plastics into different types. An infrared beam is fired into the piece of rubbish and can effectively 'see' what kind of resin the item is made from. This is clever stuff, and it's how plastic milk cartons can be picked out from PET plastic bottles (unfortunately this doesn't work with black plastic, which isn't picked up in the processing). The whole system is focused on sorting and resorting until only select, 'target' materials are left.

It's fast, with a typical modern facility processing between twenty and twenty-five tonnes of rubbish per hour along one or two different processing lines. It's also relentless; MRFs operate twenty hours a day, fifty weeks of the year at 90 per cent capacity, clocking up some 5,400 hours per annum.

This brings us to the first problem. Although the number

23. Robinson, N.: 'Cutting back packs', *Meat Trades Journal*, 28 September 2012.

of MRFs has soared over the last five to ten years, they cannot keep pace with the amount of rubbish that we now produce, particularly when it comes to plastic.

While it's ostensibly great news that we all want to recycle more plastic, the truth is that we don't have the capacity to process it. In the UK, PRFs remain rare: while we get through five million tonnes of plastic packaging a year in the UK, according to industry insiders, we only have the capacity to recycle 350,000 tonnes a year – only 7 per cent of what we use.

It would be easy to ask why we don't just build lots more fancy MRFs and PRFs, but at around £6 million a pop, many are owned and run by the waste giants, like Viridor and Veolia. Some are co-owned by local authorities under public finance agreements. These are, I'm afraid, not always the happiest of circumstances. The council just wants our bins emptied and to keep residents happy by recycling. But the waste giants must generate profit to keep their shareholders happy, and this can be tough when the price of recyclate is low. Recyclate is sold on to the global market, and this is extremely competitive. Plastic flake and pellets (which is what successfully recycled plastic eventually becomes) are in price competition with virgin oil: when the oil price drops, recycled plastic loses out because it becomes much cheaper to make new plastic from oil.

Recyclers know they have more chance of selling high-grade, clean plastics that can be sorted and processed quickly (time is definitely money in this business), and so they prioritise plastics that fit the bill as 'target material'. Often the 'target material' is clear PET – the plastic that water bottles are made from. Not only is it high-grade material, but it also is relatively uncomplicated to wash, sort and process into flakes of plastic (that remind me of soap flakes). When sold into international markets they are predominantly used to make polyester for clothing, and other food packaging.

As consumers, eager to do the right thing and buying bottles made from other bottles, there is not enough of what we call 'closed-loop' recycling. In closed-loop recycling, an old material is processed and returned to the same 'as new' state to be made into a product of the same original material. For example, if you were to unravel a jumper, wash the wool, then use it to knit an entirely new cardigan, you would be following the closed-loop model. For plastic, a bottle turning into a new bottle is the obvious example: PET plastic in water bottles can be broken down into recyclate flakes, through mechanical recycling, and used to make brand-new drinks bottles. I think of this as peer-to-peer recycling, so an item that is made of material from an item of equal value.

Most of our recycling infrastructure, however, fails to hit these dizzy heights, and settles instead for open-loop recycling. Here the PET plastic bottle, for example, is processed to form recyclate flakes, which are then used in fibre manufacturing. In other words, new material from old recycled waste. This is not as ambitious as I would like.

This might sound a peculiar distinction, after all, what does it matter what the recycled plastic flakes get turned into? But it does matter, because it's only then, in closed-loop recycling, that the recycled material holds its value and can be recycled time and time again. What we do a lot of at the moment is 'downcycling', so recovering and reprocessing material and sending it further down the recycling stream. At the moment our PET water bottles might be recycled into food packaging or polyester for clothing, both of which have low recovery rates and are less likely to be recycled next time around. In some parts of the country we've also begun to put plastic waste from agriculture – used to cover plants and sileage – into our roads. While this is better than pouring it into landfill, to me it's still a waste. We're not recovering anywhere near the value of the oil and energy that was used to make that plastic in the first place.

The target plastics might be what the recyclers really want, but they're not the only materials we are putting in our bins in the hope that they will be recycled. There are certain groups of plastic that don't rate well on the global recycling market, or that are technically difficult to reprocess. We have a particular problem with tubs, trays and pots. So while we might dutifully clean them and put them in the right bin on the right day, it is not 100 per cent certain that they will be recycled. What on earth happens to plastics that are technically recyclable, but not financially worth the effort?

Over the last decade, we've developed a side dependency on next-generation incineration, known by the industry as Energy from Waste plants (also known as waste-to-energy). These continue to be big news: recently waste giant Viridor announced that it was spending £1.2 billion on more incineration plants. Here the waste deemed 'non-recyclable' is burned to produce superheated steam, which drives a high-pressure turbine to power an electric generator. The power created is used to run the plant and the surplus is sold to the National Grid. In the period from January to April 2018, Viridor claimed it sent enough energy to the National Grid to power 330,000 homes or a city the size of Leeds for a year – the equivalent of 225 MW of electricity.

Waste-to-energy incineration has undergone what they might call in marketing a 'brand refresh'. With a legislation change in 2008, incineration was bumped up in the waste hierarchy; it would no longer be considered a means of disposal, useful for tackling 'residual' waste for which landfill was the only alternative, but plants could now be described as 'recovery' facilities. That may sound like a purely technical change, but it allowed this industry to cross the rubicon from the problem side of the picture, along with landfill, to the solutions phase, standing alongside recycling. Many communities are not so convinced. There is still unease about living next door to large incineration plants that burn plastic. There is also the issue of sustainability: once the plastic is burned, it is lost as a resource, and that is a waste.

Waste-to-energy incinerators are demanding beasts. They are an expensive investment and they need to be fed with rubbish – and that locks us into using them for the foreseeable future. This is the crux of the anxiety for a lot of environmental campaigners, including Friends of the Earth. Since its inception in 1971, Friends of the Earth has worked on waste and resources and has always taken a dim view of incineration because the process releases emissions into the environment. Based especially on CO_2 emissions, Friends of the Earth has long considered waste-to-energy plants to be 'climate damaging' technology, even the new operations. But concern also centres on the potential prioritising of incineration over recycling. It comes down to a simple choice: do we feed incinerators, or do we think smarter? As we shift our whole waste economy and start to view our plastic empties as a true resource that can be continuously recycled – something we need to do – will our investment in energy from waste plants mean that we will not have the incentive to capitalise on newer, smarter thinking?

Currently, however, we're a long way from that ideal future. Our UK recycling services are under intense pressure, and even the Energy from Waste plants wish that there was less plastic flowing into the system. There are simply not enough hours in the day to collect, sort and burn it all. And if things weren't already pressurised enough, in 2017 along came the biggest jolt to recycling in the UK that we have ever experienced; interestingly, from China.

In July 2017, the Chinese government announced a clampdown on 'foreign garbage'. To get slightly more technical, that meant bringing in tight contamination limits on twenty categories of scrap, especially waste paper and plastic. This should have rung alarm bells for us in the UK, because between 2012 and January 2018, when the limits were enforced, we shipped more than 2.7 million tonnes of plastic scrap to mainland China and Hong Kong.[24] In short,

24. Bouchier, G., Fleming, D., Ingram, M.: 'UK Recycling Industry Faces Increased Financial Stress As China Closes The Door To Plastic', *Mondaq Business Briefing*, 1 March 2018.

we used China as a giant bin for plastic waste. China was our main market, and most of the plastic waste that we generated was on its way there.

Strangely, despite this dependency and the looming embargo, we put the China problem to the back of our mind. Perhaps we thought that the Chinese authorities wouldn't go through with it. But they did. While we export our pollution from Europe, what we often tend to forget is that one day other countries will feel the same way about pollution as we do. Fast-forward to January 2018, and it became clear that the Chinese authorities meant business: China's new drive towards reducing pollution meant that it really was kicking the dirtiest recycling out. All but the cleanest bottles and other materials would be accepted, the rest would be refused entry. The ban was enforced just in time to deal with the aftermath of our biggest plastic and waste hotspot: Christmas. In December 2017, I advised readers of my eco column to recycle really carefully and to visit their household waste centres early.

In practice, you may have noticed very little difference. But look carefully, and you'll find more and more local authorities beginning to get quite fussy about what items they will take. Behind the scenes experts tell me that councils are being forced to stockpile waste plastics while they urgently look for new markets that are not so fussy about the quality of their waste imports. Those markets are likely to include Sri Lanka and Vietnam, where there is a huge lack of recycling infrastructure.

Last year I went to Sri Lanka, and saw for myself the

rubbish dumps rising up around the network of rivers near Galle, on the West Coast. These waterways teem with wildlife: on an early-morning birdwatching trip I had seen two eagles and four different types of kingfisher within the first few minutes. But for how much longer? The informal landfills are leaching toxins into the waterways. Then there's the cost to human life. Reports came through from Colombo during my trip that a 90 m mountain of waste had collapsed killing twenty-three people.[25] An Australian tourist I got chatting to said he enjoyed the country but wouldn't return; 'Too much plastic litter,' he stated matter-of-factly. As he spoke he gesticulated at the offending plastic litter, using his bottle of water as a pointer, oblivious to the fact it would probably soon be joining its brethren.

Back in the UK, there's more bad news, I'm afraid. We fought so hard to claw our way up the recycling league tables, but we may soon find ourselves slipping back down. The truth is that, when it comes to plastic, behind the scenes our national ambitions have been quietly downgraded. This is to do with cost and the intense and unrelenting lobbying by industry, manufacturers and, unfortunately, retailers. It is also to do with the difficulty of recycling plastics. Until 2016 the UK had a statutory plastic packaging recycling target of 57 per cent by 2017. Without our knowledge, that target has been reduced and our ambitions scaled down to 49 per cent for 2016.

We find ourselves in a truly peculiar situation: we're

25. 'Colombo: Death toll from rubbish dump collapse rises', Al Jazeera, 16 April 2017.

sending more plastic than ever to be recycled, but as standards are slipping (except in Wales), less is actually being recycled. In England recycling rates were down 0.7 per cent in the last year to 43 percent. We will now struggle to make a target of 57 per cent recycled waste by 2020. This target used to look achievable.

There are a variety of factors at work here, not least that cash-strapped local authorities have little money to spend on waste. Meanwhile those manufacturers and retailers who gain most by pushing plastic into our lives are getting away with taking little responsibility. Because it is me and you, through our taxes, who fund 90 per cent of the collecting, sorting and disposal or recycling of most of the plastic waste that flows into the country, while the manufacturers, brands and retailers pick up the tab for just 10 per cent. This is another complex system that needs to be reconfigured. At the moment retailers and manufacturers offset part of the packaging they push out into the world by belonging to a compliance scheme. Not only is this unfair, but it has also come to light that this was leading many businesses to overstate how much plastic packaging they actually recycled.

According to the official statistics, during 2016 our nation produced 2.26 million tonnes of plastic packaging and recycled nearly 45 per cent of that, which doesn't sound too bad.[26] But according to waste and recycling expert Dr

26. Chapman, B.: 'Plastic pollution: Businesses underestimate amount of packaging waste they produce by 50%, report finds; Recycling rate far lower than official figures suggest because of fundamental flaw in way it is worked out, meaning Britain may have missed EU targets for years', *The Independent*, 6 March 2018.

Dominic Hogg, who has investigated the real pattern of plastic recycling in the UK to produce a report for research company Eunomia, 'No one believes these figures.'[27] Oh dear.

So our humble recycling ends up in a complex system involving the price of crude oil, market forces and political personalities. Honestly, if Macbeth's witches had got together to create a system that would cause maximum bother, I don't think they could have magicked up anything more pernicious than this alliance of factors. Double, double toil and trouble.

But while this seems like a giant mess, could there be an upside? When recycling plastic works, it's brilliant. I've seen it with my own eyes, as it is sorted and cleaned in giant washing machines where the labels unstick and we're left with clear plastic that can be ground into clean lentils and flakes as good as new. There is a lot of opportunity here. If we can't dump our rubbish on China, could we finally be forced to confront the reality of our own waste habits? Can we now convert our guilt response triggered by *Blue Planet II* by taking a zero-tolerance stance on the plastic packaging problem that likely caused the whale to die in the first place? Can the plastic, manufacturing and retail industries own up to their mistakes and contribute more to the cost of the infrastructure they rely on? Could this be the point at which we demand more from our recycling while also using less?

27. Wells, L.: 'UK overestimates plastic recycling by a third, finds report', *Independent Retail News*, 6 March 2018.

A GIANT HEAP CAUSES A STIR

Fast-forward to early April 2018, and we were about to stage a major stunt that would reinforce the critical need for action. The Big Spring Beach Clean is a perennial fixture in the fight against waste. Run by the brilliant NGO Surfers Against Sewage, based in tiny St Agnes, in Cornwall, a network of pivotal beach cleans takes place every year around Easter. The Big Beach Cleans are epic battle sites of plastic-picking heroism on the part of the thousands who take part trying to repair the damage to the UK's 2,500-plus beaches. They are particularly important in giving our beaches a fighting chance approaching the summer season, when the volume of 'fugitive' plastic spikes, stemming from local sources such as takeaways, who dispense single-use 'disposable' items like coffee stirrers, sauce sachets, the dreaded plastic straw and our old friend the coffee cup. The Big Spring Beach Clean is a chance to get as much plastic as possible off the beach and to deal with it to make sure it doesn't enter, or re-enter, the marine environment. Participants collect an incredible amount: in 2017 the Big Spring Beach Clean netted 97 tonnes of plastic waste – a drop in the ocean, yes, but a significant reprieve for the local environment.

After the 2018 Big Spring Beach Clean, a proportion of that waste – 8.7 tonnes – had an extra stop-off point on its way to being sorted, recycled or incinerated. That 8.7 tonnes of plastic waste, which, 24 hours before had been collected by hand by 1,200 volunteers from 52 beaches across the UK (we wanted a representative sample), was dumped at my office – that is, right in front of BBC Broadcasting House in the heart of London's bustling West End. Our design team, wearing hazmat suits, descended on it and before long, the mound of waste was artfully arranged into an outdoor studio.

That evening, stunned viewers tuned into *The One Show* to see a dystopian scene. It was an extraordinary sight. Our entire studio had been rebuilt outside in front of the BBC from the 8.7-tonne waste heap, representing just one millionth of the plastic that enters our seas globally every year. As the cameras swooped around the set, viewers saw some of the spoils from the Big Spring Beach Clean, but they didn't get the powerful stench. It came in waves, a nauseating smell, common to landfill. I wondered if we'd be fired.

And then we were on air. 'On tonight's *One Show* we have just one thing on our minds: plastic. What more could be done to turn the tide on single-use plastic?' We dedicated our whole show to it, debating plastic with the environment secretary Michael Gove and the *Spring Watch* star and natural history presenter Chris Packham. I'm not sure that a prime-time show has been dedicated to an environmental issue in that way before, and despite the olfactory discomfort, it felt great to be part of it.

After filming ended, I had a quiet moment on the set looking at my rubbish. The team sent by recycling company Viridor were beginning to cart it away. I was soon to follow, taking part of the rubbish to Rochester and then on to Skelmersdale, where as much as humanly possible would be

Plastic number symbol & Abbr.	Name of plastic	Common domestic use
1 PET	Polyethylene terephthalate	Clear drinks bottles, oil bottles, food packaging and punnets, textiles (e.g. polyester).
2 HDPE	High-density polyethylene	Milk containers, cleaning, laundry and detergent bottles, shampoo bottles. (You can remember HDPE is the milk container plastic, because, coincidentally, it has a slightly milky look, as is opaque rather than clear).
3 PVC	Polyvinyl chloride	Detergent bottles, squash bottles, window frames, drainage pipes, shower curtains, clothing, toys, clear food packaging.
4 LDPE	Low-density polyethylene	Dry-cleaning and carrier bags, bread bags, squeezy bottles and containers, yokes for six-pack beers, linings and laminated cardboards.
5 PP	Polypropylene	Margarine tubs, soup pots, most bottle tops, waterproofing in clothes, straws, medicine bottles.
6 PS	Polystyrene	Most yoghurt pots, meat trays, takeaway cups, disposable plates, takeaway containers, compact disc holders, cushioning in packaging.
7 OTHER		Usually indicates that item is made from a blend of plastics, in any combination of the standard six plus any other resin type. Includes acrylic/perspex, nylon and polycarbonate. Certain food containers. DVDs, sunglasses.

UK recycling information
PET drinks bottles are collected by most (92%) council waste-collection schemes. Recycled into fabrics and fleece, carpets, straps. Also can be recycled into new bottles in Closed-Loop systems, and other food packaging.
Collected by most (92%) councils. Recycled into recycling containers and bins, garden furniture, pipes, pens. New technology allows HDPE to be recycled for new milk bottles.
Not generally collected in household recycling. Can be hazardous for recycling centres to deal with.
Not generally collected in household recycling. Plastic carrier bags are collected by some supermarkets for recycling into new carrier bags or bin bags. Mixed plastic recycling is hoped to be in place within five years.
Not generally collected in household recycling in spite of potential – for recycled plastics for bins, pallets, trays etc. Again, mixed plastic recycling is hoped to be in place within five years.
Not generally collected in household recycling. Some commercial polystyrene may be recycled however for use in foam packing, rulers, carry-out containers.
Not generally collected in household recycling.

turned into clean flakes of PET (the most common sort of plastic, used in water bottles) ready to make new stuff. It was a lot to process, both emotionally and from a recycling point of view, and I felt a strange elation that the issue we had been campaigning and making films about for so many years was suddenly at the forefront of everyone's minds. Plastic from the ocean was laid out before my eyes. But why on earth was it there in the first place? And what, or who, was to blame?

4

TRASHING THE PLANET

The amount of plastic debris in the sea is predicted to increase from 50 million metric tons in 2015 to 150 million metric tons by 2025.[28]

Grappling with the plastic in your life means getting your head around the flow of this material into and around the natural environment. Look at the everyday items around you: a sandwich wrapper or a takeaway carton made from blown polystyrene; a tub of chewing gum; a pair of sports socks; or even the common-or-garden teabag. All depend on plastic for their manufacture. Each time you make a cup of tea, or chew gum, or throw away a chocolate wrapper, you probably don't automatically think it may lead to ocean gyres, or the death of a whale, or enter the bellies of zooplankton – but read on. Plastic's odyssey is like no other.

28. The *Future of the Sea* report, UK Government, 2018.

I spend a lot of time going up and down this beautiful island, travelling to far-flung locations. The Victorian pier and beach at Clevedon on the outskirts of Bristol is a popular place for filming. It is secluded and genteel, and not too busy out of season. On the drive from Portishead, beautiful postcard villages are reached along roads edged by generous grass verges. But you can't help noticing the litter: plastic bottles and cans scattered among the wildflowers, hedges decorated with crisp packets – the usual problem items in problem places. I immediately had vengeful thoughts about the litter-louts who'd left it there, but something didn't quite fit: the upper canopy of the trees was garlanded with plastic supermarket bags. These were in a raggedy state; given that the bag tax has caused a huge drop-off in usage, they'd probably been there for a while.

Committed litterers, local fly tipper, or something else? I rounded the corner and there it was: the sign to a waste transfer site where rubbish was trucked in several times a day. Inevitably, given the volume and pace of the waste traffic, windblown escapee rubbish is the most likely culprit. It was a reminder that this can't and shouldn't always be blamed on rogue litterers.

Raised on a rubbish diet of Keep Britain Tidy and a commendable sense that putting your junk in the bin is a civic duty, the UK has a particular obsession with the 'litter-lout'. As the subject of plastic pollution comes up more and more, it takes approximately forty seconds in my experience for someone to mention litter-louts and how much they hate them. I'd like this to be taken as a given. Can we just agree that yes, they are awful? I mean, who in their right mind would applaud someone who chucked fast-food wrappers out of their car window? Case closed.

What I'd love to consider instead is whether the litter-lout, the bogeyman of the plastic pandemic, isn't actually a wildly overblown construct, a convenient semi-truth that's let the real culprits in the waste pandemic off the hook. Controversial, I know.

Of course, I've seen the odd person drop litter, and don't get me started on fly-tipping. We bemoan declining standards of behaviour, and we even have a suspect in our sights: men drop three times as much litter as women,[29] and the young, aged between sixteen and twenty-four, drop twice as much as everyone else. It's easy to blame the litter-lout, and who doesn't love a handy scapegoat?

In England, we're heavily invested in flushing out the litter-lout. There's a £500,000 Litter Innovation Fund, for example. The focus is traditional: change the behaviour of litter louts and galvanise all of us, including the hardcore recycling refuseniks, into packs of litter-picking volunteers. One report from the Keep Britain Tidy Centre for Social Innovation caught my eye.[30] 'We're Watching You' is essentially a guide to how to stake out the bins at Beaconsfield Motorway Services! A graphic illustrating a pair of beady eyes in a rear-view mirror accompanies handy tips on observing behaviours and intercepting litterers. If we are

29. 'Keep Britain Tidy', *Journal of Litter and Environmental Quality*, Volume 1, Number 1, June 2017.

30. Case Study: We're Watching You – Beaconsfield Motorway Service Area, March 2015. keepbritaintidy.org/research

to believe the Keep Britain Tidy literature, eradicating litter is merely a question of education and surveillance. Hmm, I'm not so sure.

Dr Sherilyn MacGregor from Manchester University has studied the government's litter strategy.[31] Her research centres on Moss Side, an area of Manchester with 'a large population of students studying at universities with award-winning sustainability education programmes', and one also full of alleys strewn with rubbish dumped by students. Education, MacGregor concludes, does not seem to be the issue here; rather local government funding cuts that have purged the street cleaners and sweepers and left litter to proliferate, this along with a policy that allows the manufacturers of single-use packaging to get off scot-free. Meanwhile, responsibility for cleaning up is dumped on un-incentivised volunteer communities. We can surely do better than this?

Again I return to Keep Britain Tidy to take a look at the businesses that generously sponsor their activities. Among the corporate partners in 2018 for their annual Great British Spring Clean are Coca-Cola, Costa, McDonald's, Lidl and DS Smith, the leading packaging company in the UK which last year posted profits of £4.7 billion. The fact is that the growth in litter – up 500 per cent since the 1960s – mirrors almost exactly the growth of the packaging industry, particularly of the single-use container. Yet up until now the brand and packaging giants take little, if any, of the flack.

31. MacGregor, S.: 'Why England's new litter strategy is actually a bit rubbish', *The Conversation*, 14 September 2017.

I have a proposal: that we spend less time on surveillance and scanning for litter-louts, and channel our time and energy into reducing plastic at source, before it turns fugitive.

Once fugitive plastic is on the move, it's difficult to stop. An enormous 8 to 12.2 million tonnes (depending on which data sets you look at) of plastic waste ends up in the marine environment each year.[32] Rivers are a major source of plastic pollution, delivering bottles, stirrers and coffee cups with incredible regularity. In fact, they are thought to deposit between 1.15 and 2.41 million tonnes of plastic in the sea.[33]

Meanwhile, some stuff seems supranaturally clever at hurling itself into the water – 10 per cent of the litter in the River Thames, for example, is made up of plastic bottles.

Consider the weight of an empty plastic bottle and how easily it rolls and moves, and you get a clearer sense of how an estimated 80 per cent of marine litter comes from the land itself.[34] When you live on a wet and windy island like we do, any plastic waste that isn't properly disposed of, bagged and tied down will be whipped up by the wind and rain and will travel in a seaward direction. This litter is a sign of a system that cannot cope. An overflowing bin in a beach car park is a warning sign. Once it's in the water it's difficult to collect, despite the heroic efforts of beach cleaners, and then it gets swept out to sea.

32. Sherrington, C., Dr: 'Plastics In the Marine Environment', *Eunomia*, June 2016.

33. Lebreton, L. C. M., van der Zwet, J., Damsteeg, J.-W., Slat, B., Andrady, A., Reisser, J.: 'River plastic emissions to the world's oceans'. *Nature Communications*. 2017; 8:15611. doi:10.1038/ncomms15611

34. Kinsey, S., MCS: 'Scale of Plastic Bottle Waste', written evidence to the Environmental Audit Committee, UK Parliament.

GYRES AND GARBAGE PATCHES

If overflowing bins and waste trucks don't capture the imagination, what happens to some of the trash that finds its way into the world's oceans is mind-boggling – but for all the wrong reasons. Once it enters into the marine environment, the movement of the swirling vortex of plastic is more dynamic, more damaging and more peculiar than anything we could imagine.

'You must be able to lift one-third of your body weight,' reads an intriguing advert that appeared in a number of US West Coast news organisations in 2011. Part environmental message, part enticement to pay up and take a new form of eco-cruise, it represents an early example of 'garbage patch' tourism. As well as strong and physically fit, you also needed to be pretty wealthy: $10,000 would buy participants a chance to spot swirling plastic detritus and see the promised scenery such as cigarette lighters, bottle caps and toys churning in a vast plastic whirlpool, or gyre, that has become known as the Great Pacific Garbage Patch (GPGP). Gyres occur when airflows moving from the tropics to the polar regions create a clockwise rotating air mass, which then drives oceanic surface currents in the same direction. This is what we have discovered about the movement and behaviour of plastic trash out at sea – once the detritus enters the ocean currents, the buoyant plastic is inclined to settle in islands of trash that float just above the surface. It is here, where winds are light, that the plastic debris of our throwaway lives is dramatically visible.

To see the particularly spectacular gyre on the 2011 expedition, you also needed a sizeable chunk of annual leave. The cruise, aboard a 72-foot racing sloop, would take 20 days sailing 4,490 km across the Pacific from Honolulu, Hawaii to Vancouver, British Columbia.

This 'science' cruise was actually a fundraiser for Algalita, a non-profit organisation set up by yachtsman and oceanographer Captain Charles Moore, who coined the name the Great Pacific Garbage Patch for the rubbish-infested North Pacific gyre. He stumbled across the GPGP while he was captaining a racing yacht in 1997. I can't imagine how he must have felt to find himself in a vast sea of floating debris as far as the eye could see. While his eyes might have had trouble adjusting to this new reality, it provided the rest of the world with evidence: there was a price to pay for bingeing on plastics, and it was being paid principally by the world's oceans.

Talking about the way human waste congregates in these gruesome gyres, my friend Liz suddenly remembered her grandfather, a distinguished colonel, telling her about seeing floating trash. Now ninety years old, we gave him a call on her mobile. Colonel John Weston confirmed that he was stationed at the Hawaiian island of Oahu in 1976, and that when out at sea he spotted an extraordinary amount of plastic. I like to think he saw the GPGP two decades before it was given a name. Certainly the colonel's recollections demonstrate that the plastic plague was already forming in the 1970s, within twenty years of plastic materials coming into everyday use. It was an extraordinarily rapid rise, with an equally rapid impact. Forty years on, Oahu has a serious problem with plastic waste, the island acting as a net for the Great Pacific Garbage Patch which periodically spills its contents on the shore.

The GPGP has grown at a shocking rate. The most recent team that went in search of it was made up of scientists from seven countries brought together by the Ocean Cleanup Foundation. In the tradition of ocean plastic science, the team used the tested technique of towing fine-mesh nets behind the boats[35] to pick up surface samples. This study was supersized and supercharged. It sent thirty towing vessels crossing the GPGP to collect samples. Meanwhile the project's mother ship RV *Ocean Starr* trawled two six-metre-wide devices to pick up the medium- to large-sized objects excluded from conventional net tows. Flying above on the tail of the trawler vessels, a C-130 Hercules aircraft fitted with advanced sensors recorded and collected multispectral imagery and 3D scans of the samples as they were found.[36]

The team of researchers found that the GPGP is sixteen times bigger than we thought from previous estimates. Stretching across 600,000 square miles of ocean, it dwarfs France, is bigger than Texas, weighs in at 79,000 tonnes and

35. I checked with several marine scientists and they confirmed that yes, their nets are made of plastic, which shows how we're all co-opted into using plastic at some point. An irony, given this study found that at least 46 per cent of the found trash came from fishing nets.

36. Lebreton, L., Slat, B., Sainte-Rose, J., Aitken, R., Marthouse, R., Hajbane, S., Cunsolo, S., Schwarz, A., Levivier, A., Noble, K., Debeljak, P., Maral, H., Schoeneich-Argent, R., Brambini, R., Reisser, J.: 'Evidence that the Great Pacific Garbage Patch is rapidly accumulating plastic', Scientific Reports – *Nature*, 8, Article No. 4666, published 22 March 2018.

contains an estimated 1.8 trillion pieces of rubbish, 99.9 per cent of which is plastic. One item pulled from the patch was found to be forty years old. It could have been bobbing around when Colonel Weston made his trip.

A GIANT SMALL PROBLEM ...

The ocean gyres, particularly the Great Pacific Garbage Patch, have taken on quasi-mythical status. Any young adventurer worth their salt is desperate to get out there and start vlogging for *National Geographic*. But while everyone was freaking out about the size of these gyres and the amount of plastic in them, back at the University of Plymouth, Professor Richard Thompson was asking a different question altogether. He looked at the rapid rise in plastic usage, the huge amount 'stored' in gyres and he wanted to know where the rest of the plastic was.

To answer this question, day after day Thompson began to collect his own data. Rather than getting a speedboat out to some gorgeous reef in Mauritius or to the Baa Atoll in the Maldives, Thompson and his team pulled on their waders and trudged out into the estuary of the English Channel near to their laboratory in Plymouth, and there they literally sieved sediment. This is no mean feat. What Thompson and his colleagues found were tiny bits of plastic, each under 5 mm. He called them 'microplastics', a term that we're beginning to get increasingly used to hearing.

In the fight against plastics, we've become obsessed with

microplastics, and with good reason. Remember that 80 per cent of plastic in the ocean originally comes from the land, and likewise most of these tiny microplastic fragments began as recognisable objects, rubbish that we didn't quite dispose of properly or that escaped the refuse system and ended up in the sea. Once washed into the tides, these plastic objects – a bottle or crisp bag, for example are tossed and churned by the waves as if in a never-ending laundry cycle. The macro plastics are constantly tumbled and thrown up against the shingle and other abrasive objects on the shore, and are broken down by UV rays until the objects get smaller and smaller, becoming minuscule microplastics. And these pollutant fragments constitute a real risk to marine life.

They can also be easily mistaken for food. In the main, microplastics form a toxic soup, suspended below the surface of the seawater creating an effect described by divers as ocean smog.

I visited the laboratory at Exeter University where scientist Dr Matt Cole invited me to view copepods under the microscope. Copepods are a group of small crustaceans found in both the sea and in freshwater habitats and are one of the primary ingredients of zooplankton, the collective term for microscopic organisms that drift with water currents and on which almost all oceanic organisms are dependent as a food source. Dr Cole's research is on the biological and ecological impacts of microplastics in the marine environment. Here he was studying the effect of microbeads, a manufactured microplastic that, until a ban was introduced via legislation, was a regular ingredient in cosmetics and personal care products, especially in skin exfoliators. In Matt Cole's laboratory the microbead is given a neon marker so that he can track it under the microscope as the copepods ingest it.

When living organisms ingest microbeads and other microplastics, mistaking the plastic particles for food, their energy levels are depleted as plastic offers no sustenance. As a result these organisms may die before they reproduce, interrupting the life cycle of the species. Some studies show that the reproduction of oysters and crabs living in microplastic-saturated water is halved, for instance.[37] And, because zooplankton (which may have ingested microplastic) is a food source for a multitude of ocean creatures, including whales, this is one of the most insidious ways in which the plastic we throw away enters the food chain. And enter the food chain they have.

As if this wasn't concern enough, a new, additional plastic peril has recently been identified. Australia-based scientist Mark Browne has fought long and hard to get us to recognise a new environmental danger emanating from our growing addiction to plastic-based textiles – microfibres. Over the previous decades our wardrobes have shifted from natural fibres such as cotton, wool and silk to synthetic man-made fibres – today many of us live in techno-fibre sports apparel as if at any minute we might be called upon to run a triathlon. His research, beginning some ten years ago,

37. Gall, S.C., Thompson, R.C.: 'The impact of debris on marine life', *Marine Pollution Bulletin*, 15 March 2015; 92(1–2):170–179.

discovered that the majority of shoreline microplastics were actually microfibres from textiles.

Meanwhile, our old friends the marine team at Plymouth University have found that during the wash cycle in a normal washing machine, an acrylic garment can shed upwards of 700,000 microfibres as the fibres escape through the rinse and drain cycle to become another source of microplastics.[38] And although acrylic fibres that include fake fur shed five times more than polyester-cotton blend fabric, and 1.5 times as many fibres as polyester, the research shows that all synthetic fibres are a source of more microplastics. Flushed into our drains through millions of washing machines during millions of daily wash cycles, these plastic microfibres eventually find their way into our rivers and then into our seas and oceans.

At the moment, only 30 per cent of the world's population have access to washing machines. The other 70 per cent, however, would probably like them. As developing economies emerge, a consumer class will undoubtedly want to forgo the hand-washing of clothes. We'd better pray they don't also want to wear synthetic fibres. This is the laundry plastic pandemic that we are yet to square up to.

38. Napper, I. E., and Thompson, R. C.: 'Release of synthetic microplastic plastic fibres from domestic washing machines: Effects of fabric type and washing conditions'. *Marine Pollution Bulletin*, 2016, *112* (1), 39–45.

PLASTIC SPILLS

There's another ready-made microplastic pouring into waterways that comes directly from the plastic industry. I went to South Devon to meet Marion, an experienced beach cleaner. As we began our clean-up down on the shore, Marion beckoned me over and funnelled some little pellets from her hand to mine. Evidently she is much sharper-eyed than me – I hadn't noticed these tiny plastic granules of differing shapes, about the size of a peppercorn or a lentil, as I'd been scouring the sandy ground. Known as 'nurdles', the pellets in the palm of my hand weighed next to nothing: some white, others multicoloured, some opaque, others translucent as a dewdrop.

To surfers and beach cleaners, these are known by the poetic name of 'mermaid's tears'. Once you're alert to the nurdles and schooled in zeroing in along the shoreline, it's quite satisfying to pick them out from the bladderwrack (another rather poetic word) seaweed strewn across the shoreline. The origins of nurdles are more prosaic: they are the raw material and minute building blocks of almost every bit of plastic we consume from PVC to cling film.

Marion is what I would affectionately call a 'nurdle-chaser'. She honed her skills in spotting and picking them up along the banks of the river Clyde where they're particularly numerous, transported by currents from Strathclyde's plastic manufacturing plants nearby. Any time you're transporting tiny bits of plastic, there's a risk of creating plastic waste and occasionally millions of the pellets will spill from an overturned container. Marion shows me a nurdle app on her phone and then she adds the date and information on the location of the nurdles she has just collected. The app data feeds into a database which tracks sightings of nurdles across the UK and can be used to map their trajectory to discover how they end up on UK beaches.[39] Beachcombers and nurdle-chasers like Marion are not just cleaning up beach litter, but performing so-called citizen science. Ultimately, their quest is to pinpoint the likely source of the nurdles they find. They are collecting evidence. Escapee nurdles are a direct link between production of the material and the plastic pandemic. While the plastic industry has taken steps to prevent spillages, it can't contain all of these tiny bits of plastic. The fact of the matter is that the more plastic production, the more nurdles escape. If the industry didn't generate so much new plastic, then fewer nurdles would end up being created and transported.

Like the microbeads used in beauty and cleansing products, these small but deadly pollutants don't have to degrade to pose a threat: they are a ready-made microplastic. And again, like microbeads and microplastics, unsurprisingly nurdles are not good news for us, or for the environment and wildlife. While they are obviously small, microplastics have a huge relative surface area to which toxic chemical substances can attach themselves, meaning that wherever the microplastics travel, Persistent Organic Pollutants (POPs) travel with

39. https://www.nurdlehunt.org.uk/take-part/nurdle-map.html

them. These are a particularly worrisome class of chemicals and include DDT, an organochloride chemical originally used as an insecticide. POPs are toxic to both humans and wildlife, and on land humans have taken big steps to clean up and phase them out altogether. To many researchers, the fact that microplastics can concentrate them and have the potential to carry them far and wide into the food chain represents a nightmare scenario.

Increasingly microplastics are entering our bodies too. The average European seafood eater is thought to ingest 11,000 pieces of microplastic a year. Given that plastic and microplastics have accelerated in the last ten years, and allowing time off for early childhood when – aside from fish fingers – I was not a great seafood consumer, at the ripe old age of forty-three I probably have at least thirty years of serious fish and seafood eating under my belt. That means I could have ingested 330,000 fragments of microplastic in my diet. And that's only the microplastic in the seafood or fish I may have eaten – on average, 83 per cent of drinking water samples are contaminated with them, and they have been found in salt, beer and honey.

GRIM DISCOVERIES

The impact of plastic on our health is still being determined. I don't mean to be flippant, but in a way, we're the lucky ones. Whatever you think of my table manners, I do not suck in my prey, nor do I have four stomachs (on a good day).

The sperm whale is not so fortunate. It is particularly vulnerable to plastic pollution, as we saw from *Blue Planet II*. We know that it takes in the region of 29 kg of plastic to kill a sperm whale because in February 2018 a six-ton, 33-foot-long juvenile male sperm whale beached near a lighthouse in Cabo de Palos, in Murcia, Spain. In April the results of the necropsy were released, revealing the gruesome 29 kg statistic: plastic bags, pieces of net and a plastic jerrycan were pulled from the animal's four stomachs, tagged and weighed.

Necropsies, the animal equivalent of an autopsy, are becoming regular occurrences as more vulnerable sea creatures succumb to death by trash. A disturbing photograph from back in 2011 shows Dr Alexandros Frantzis, Scientific Director at the Pelagos Cetacean Research Institute in Greece crouched next to a hundred plastic bags and other pieces of plastic debris that he had recently pulled from the stomachs of another deceased sperm whale found off the Greek island of Mykonos. To put it bluntly, it looks like a murder scene. When the blood was cleaned from one of the plastic bags, it displayed the phone number of a restaurant in Thessaloniki. This image helped to focus attention on the need for a tax or ban on plastic carrier bags.[40] But it would take another seven years until a four euro cent charge on carrier bags was introduced in Greece in January 2018, causing usage to drop by 80 per cent in the first month.

More recently, the grim findings from the necropsies

40. Kraft, C. X.: 'Sperm Whale: Death by 100 Plastic Bags', 7 June 2012, oceanwildthings.com

of thirteen of the twenty-nine whales that beached in the German province of Schleswig-Holstein were made public. Amid the plastic debris found within the cadavers' intestines were a 13 m fishing net, a 70 cm piece of plastic from a car engine cover and a plastic bucket. At a sombre press conference, the German environment minister suggested that the animals may have thought the items were food, mistaking plastic for squid. The animals starve, thinking they have full stomachs.

There are fewer and fewer good news stories about whales and other aquatic wildlife hitting the headlines. Meanwhile experts warn that we shouldn't just be worried about large plastic objects such as the jerrycan or entire fishing nets blocking the gut, but the small bits of microplastic, the nurdles and the microfibres, too, which have the potential to harm all species of cetacean – dolphins, whales and porpoises – not only those that suction-feed.

Over 280 species of wildlife including puffins and fulmars have now been found to ingest microplastics. In March 2018 a study reported in the *Frontiers in Marine Science* journal revealed that three-quarters of deep-sea fish have plastic in their stomachs.

It's hard to express adequately how catastrophic this is. It is difficult to process the fact that whales, sentient creatures with the largest brains of any animal that's ever inhabited the earth, are being killed due to our discarded plastic debris, the fallout from our mindless 'throwaway' acts of consumption.

5

THE FOOTPRINT OF PLASTIC

FIGHT THEM ON THE BEACHES

The endgame of this giant plastic binge is revealed in all its nightmarish glory on our beaches. 'Trashed tidelines' are the evidence of plastic's ability to travel, endure and fragment. Of the eight million individual pieces of waste that are reckoned to enter the world's marine environment every day, most are made of plastic. Around 70 per cent of the rubbish sinks to the seabed, 15 per cent drifts upwards in the water column and 15 per cent is deposited on our shores.[41] Living in Europe, the world's largest maritime zone, and as islanders to boot, we can't avoid the spectre of ocean trash.

In the early 1990s the Marine Conservation Society began to tackle the issue of waste on our UK beaches by organising

41. 'Marine Litter in Europe Seas: Social Awareness and Co-Responsibility', D1.1, Review of the Current State of Understanding of the Distribution, Quantities and Types of Marine Litter, EU, July 2013.

121 volunteer 'beach watches' across the UK. An advert in a local newspaper for the Easter beach clean at Hart Warren in Cleveland in 1993 promised sewage, plastic and metal debris. So, quite the variety. The 2018 Big Spring Beach Cleans organised by Surfers Against Sewage mobilised 35,500 people across 571 beaches to collect 63 tonnes of plastic waste. It's an incredible number, but every person was needed. Plastic litter has increased by 140 per cent since 1994, and over the past fifteen years the amount of litter washing up on British beaches has doubled.[42] This was borne out by the huge haul at Sennen Cove, where I joined the volunteers to carefully pick plastic strips, microbeads and fragments from between the rocks.

The world's largest, and longest-running, volunteer beach-clean operation is reckoned to be at Versova beach,[43] a one-and-a-half-mile strip of coastline facing the Arabian Sea in western Mumbai. The area, once the centre of a thriving, age-old fishing culture, had become synonymous with plastic waste. In October 2015, local lawyer, activist and ocean lover Afroz Shah decided to start his own beach clean initiative and quickly built an army of fellow volunteers. By March 2018 it was reported that the volunteers had hand-collected an astonishing 13 million kg of plastic rubbish, and at least eighty hatchling turtles of the vulnerable Olive Ridley species were spotted on the beach for the first time in decades.[44]

42. 'Marine Litter Report', Surfers Against Sewage, October 2014.

43. Harvey, C.: 'The world's largest beach clean up has cleared more than 4 million pounds of trash', *Washington Post*, 15 August 2016.

44. Safi, Michael: 'Mumbai beach goes from dump to turtle hatchery in two years', *The Guardian*, 30 March 2018.

Over many years, beach cleaners around the world have painstakingly, quietly and sometimes not so quietly cleaned up and collected evidence of the plastic pandemic. The hauls have been collated and analysed, and the data, with its shocking figures, represents powerful, citizen-led scientific proof in the campaign to stem the tide.

Thanks to beach clean data, we also have a clearer idea of which items are entering the ocean in the greatest numbers. The Ocean Conservancy (which has mobilised over twelve million people on global beach cleans) produces its top finds from beach cleans annually. Seven out of the top ten items listed (in reverse order) are plastic:

- plastic lids
- metal bottle caps
- plastic grocery bags
- glass bottles
- plastic bags (non-grocery)
- straws and stirrers
- plastic bottle caps
- food wrappers (including sweet wrappers)
- plastic drinks bottles
- cigarette butts

The incontrovertible evidence collected through beach cleans is a powerful weapon in the fightback – without it, I doubt we'd be on the brink of change today. Politicians tend to take notice when large groups of people mobilise for a cause. If you haven't been on a beach clean, what's stopping you? (see Further Information, details for Surfers Against Sewage who run beach cleans, page 255).

FOOTPRINTING PLASTIC

We've seen which plastic items are the most persistent polluters in the marine environment: now let's take a closer look at those which are probably invading your own life. Living in Western Europe, in a heavily industrialised society, makes us de facto heavy consumers of every material, not just plastic. We are encouraged to consume, and we're very good at it. According to parliamentary research, British household consumption remains the 'main engine of growth for the UK economy', accounting for 63 per cent of GDP: if we stop consuming, the economy tanks – or at least, that's what we're told.[45]

Now, there are many things I like about being a privileged consumer. I would rather be on this side of the fence than one of the 783 million who live in 'extreme poverty', classified by the World Bank as living on below $1.90 a day.[46] Being on this side of the fence also affords me a certain amount of power to change things; I can direct my spending towards the services, products and systems that are more just and ecologically balanced, products that display the Fairtrade

45. 'Components of GDP: Key Economic Indicators', Office for National Statistics, UK Parliament Publication, 25 May 2018.

46. Extreme poverty concentrated in sub-Saharan Africa https://data.worldbank.org/indicator/SI.POV.DDAY?locations=ZJ-8S-Z4-Z7-ZQ-ZG&start=2013&end=2013&view=bar

stamp, for example. But while empowered, I am also acutely conscious that, in the main, it's the planet that is picking up the real bill for the stuff I enjoy.

Our ecological debt is enormous. Everything we consume depletes the earth's resources. In the early 1990s Mathis Wackernagel, an engineering student at the University of British Columbia in Canada, decided to account for all the stuff we apparently need to support our lifestyles. This included grazing land for cows for meat and dairy, forests for paper and wood products, water for everything. Then he calculated the resources used in a year by an average individual, and compared this with the earth's capacity for regeneration, i.e. the resources that the earth can replenish through natural cycles during the year. Wackernagel's calculations showed that consumers in Western Europe were deeply in the red, burning through resources faster than the earth can replenish. If everyone on the planet were to consume at the rate we do in the UK, we would need three whole planets' worth of resources to support us.

Wackernagel called his model of accounting 'ecological footprinting', and it transformed the thinking on consumption. We are now well used to thinking of our environmental and carbon footprints, even if we don't do it very often. (Calculate your ecological footprint at footprint.wwf.org.uk.) The annual Earth Overshoot Day is held on the day that statistically we exceed the earth's capacity to regenerate life-sustaining resources. It's not exactly cause for celebration; in 2017 it fell on 2 August, and every year it creeps forward by a few days. The goal is to push it back.

YOUR PLASTIC PROFILE

Evidence garnered from the beaches might be frightening and depressing, but it has served to crystallise the problem of how we deal with our rubbish, and to demand that we reappraise our waste and recycling industry to see where the leaks have occurred. It has also spurred us into action and provided a kick up the proverbial to policymakers, institutions and think tanks, inciting them to drill down and release reports on plastic consumption in the UK. This is long overdue, because when it comes to plastic, every day is pretty much Earth Overshoot Day.

I've used the latest think tank data to form an approximate plastic profile of the average UK consumer. And although my picture is, by necessity, based on estimates and generalisations, it is instructive – not least because it reveals some disturbing trends in our consumption.

Our plastic footprint in the UK is around 139–140 kg per person, per year.[47] That is *three times* the amount of plastic per person that we consumed in 1980,[48] and the equivalent of 11,024, average-weight, empty PET water bottles (at 12.7 g each). Of course, if our personal plastic consumption were entirely to consist of water bottles, it would make things easier – we're quite good at recycling PET bottles, at least if we get them in the right bin. But our plastic use is much

47. Van Sebille, E., Dr Spathi, C., Dr Gilbert, A.: 'The ocean plastic pollution challenge: towards solutions in the UK', Grantham Institute briefing paper no. 19, July 2016.
48. Van Sebille, E., et al.: 'The ocean plastic pollution challenge'.

more complicated than that. To drill down into the plastic profile, here are the numbers on a few of the most common household plastics:

Water bottles You will likely plough through 150 plastic water bottles every year. In London, usage rises to 175 plastic water bottles per person, per year.

Cling film This is made from thin, stretchy plastic, polymer plastic #3, which is problematic in recycling. We are a nation of devotees: in the UK we get through 1.2 billion m of cling film every year, enough to wrap the world thirty times.[49] Your household's share is 44 m per year.[50]

Toothpaste tubes You will each get through a cool (and minty) eight to ten average-size tubes of toothpaste a year – or 21.5 tubes if you are a fan of travel-size packs.[51] Most of these are now plastic, and you will struggle to find a council that will accept them for recycling.

DISPOSABLE SINGLE-USE PLASTIC PRODUCTS

The UK ranks fifth in the EU's single-use plastic consumption chart of shame. The smoking ban, a change in cigarette advertising and education have engineered a big shift: we

49. Plamondon, C., Sinha, J.: 'Life Without Plastic', extract *Daily Mail*, 6 February 2018.
50. Divided by 27.1 million households, number of households in the UK Office for National Statistics.
51. Working on the assumption that UK citizens use a similar amount of toothpaste to Americans, given healthcare routines and marketing.

have fewer smokers, therefore fewer cigarette butts which contain plastic, and this brings down our overall UK total. If it wasn't for this fact, we'd be in the top two: we consume more cotton buds and sanitary towels containing plastics than any other European nation and we're among the biggest users of straws on the planet. Saved from total disgrace by the cigarette butt, who'd have thought it!

Each year in the UK we get through:[52]

- **Drinks cups and lids** 4.1 billion single-use paper cups (part plastic; polymer-coated)
- **Plastic straws** 42 billion per year
- **Wet wipes** 10.8 billion
- **Cotton buds** 13.2 billion
- **Disposable plastic cutlery** 16.5 billion pieces
- **Sanitary towels** 4.1 billion

OTHER PLASTICS

Packaging The lion's share of current UK plastic use, packaging accounts for 67 per cent of plastic that is chucked out. Nearly 40 per cent of the plastic we put in our bins could, or should, have been recycled, but as we know, there are limitations on – and confusion among consumers about – our recycling system. The healthy recycling statistics posted

52. Elliot T., Elliot L.: 'Plastics Consumption and Waste Management, Final Report', *Eunomia*, 26 March 2018.

by the plastics industry have been called into question. In fact, waste experts at Eunomia, a UK recycling and resource efficiency consultancy, didn't question as much as trash them. 'It seems reasonable to state that no one really knows what the real recycling rate for plastic packaging currently is,' says a recent Eunomia report.[53] Ouch. Sadly, you also have to assume that some householders just cannot be bothered.

Toys Out of sixteen of the main types of consumer goods, from food to automobiles, the toy sector is the most plastic-intensive. From every $1 million revenue a toy brand makes, it uses 40 tonnes of plastic.[54] If the toy industry had to pay the true environmental cost of producing, collecting and cleaning up plastic, it would go bankrupt. And if you've have a busy Christmas with lots of young children at home, your plastic profile will likely balloon.

Agri-plastics This is the film that covers fruit and vegetables, or the containers and bags for fertilisers. Used both behind the scenes and front-of-house in the tobacco and food supply chain, agri-plastics constitute up to 1.9 million tonnes of plastic every year in the EU.

53. Hogg, Dominic, Dr: 'Plastic Packaging, Shedding Light on the UK Data', *Eumonia*, March 2018.
54. UNEP (2014): 'Valuing Plastics: The Business Case for Measuring, Managing and Disclosing Plastic Use in the Consumer Goods Industry', Copyright © United Nations Environment Programme (UNEP), 2014.

THE PLASTIC DISCLOSURE PROJECT

These figures can only ever be a snapshot and a rough estimation of our likely average plastic consumption. Piecing together our personal plastic footprint serves to highlight a major issue: the lack of transparency from big business and the plastic industry itself. The corporations that rely on plastic can be very shy when it comes to talking numbers. Coca-Cola, for example, to name just one of the big global brands, wasn't keen to disclose the number of plastic bottles it uses every year, but in 2017 Greenpeace was able to estimate reasonably that it was in the region of an astonishing 110 billion a year.[55] The NGO was also able to establish that Coke had increased production by an extra billion bottles over the year.

The global giants must, however, be compelled to come clean. Knowing the figures isn't just about arbitrarily naming and shaming global brands; it's about understanding what we do next and creating the platform for action. 'If you don't measure it, you can't manage it' is the strapline of the Plastic Disclosure Project, established in 2010 by concerned ocean scientists. With the United Nations Environmental Programme (UNEP), the Plastic Disclosure Project has repeatedly warned major brands that they must do more to calculate and reveal the amount of plastic they use, and commit to reducing their usage year by year.

55. Laville, S.: 'Coca-Cola increased its production of plastic bottles by a billion last year, says Greenpeace', *The Guardian*, 2 October 2017.

As long as people like us keep up the scrutiny, continue to make noise and make it clear that an abundance of plastic is a big issue for us, then more brands will engage with the Plastic Disclosure Project and make public their usage. The world's corporations know that the alternative to a voluntary initiative like this is tough legislation, including bans on disposable items. In fact, we're already starting to see this: the UK government recently muted a ban on face wipes (I explain why it's necessary to wean yourself off the wipe in chapter 13, Rethink, on page 189).

I'm in favour of bans, to a point. The issue with bans is that they tackle one item at a time. And without changing the culture, and our desire to use disposable products, as soon as you've tackled one item and removed it off the list of the top ten offenders, the next just moves up and takes its place. As I know from my early interaction with supermarkets on removing plastic packaging at the behest of my readers, when I lobbied them about shrink-wrapped coconuts and secured a victory, only to find myself lobbying for the de-plastification of the cucumber a couple of months later . . . and so it continued. I call it the Coconut Conundrum.

You can get an idea of your basic plastic footprint using a calculator provided by Greenpeace: https://secure.greenpeace.org.uk/page/content/plastics-calculator. But neither footprint nor profile is as precise as we need. Unlike Wackernagel's very smart accounting system, I can't tell you how much plastic affects the planet and what our fair share of that is. All we can say is that the impact is serious and growing. As a society, we must make some crucial decisions about the biosphere. The Paris Climate Agreement, for example, gives us goals and a timeline for cutting our greenhouse gas emissions. We have a chance to avoid catastrophic climate change if we meet those goals on deadline. Similarly we need to make decisions about plastic. We can all pretty much agree that we can't continue to use it at this volume. So how much will we use, and where? On what terms? And if we decide some plastic is necessary, what would our fair share of plastic look like?

Our UK plastic profile is currently broad-brush and based on estimates: to get a real grip on personal consumption, we need better access to data. In chapter 7, Record, on page 121, I suggest how you can keep a plastic diary to get a truer picture of your plastic consumption and use the information to wean yourself off as much unnecessary plastic as possible.

THE EVERYDAY PLASTIC PROJECT

On a September evening in 2016, Daniel Webb was confronted by plastic waste that had been swept in by a storm, strewn along Margate seafront on the south coast of England. Angered by the sight, Daniel became intrigued by the fate of his discarded household plastic. As a fairly reliable recycler, he had assumed it was being taken care of, but could he be responsible for some of the plastic littering his local beach that day? When he began to investigate the fate of his plastic waste, once the bin men took it away, he found official statistics lacking and data from the recycling companies patchy.

Daniel describes himself as an everyday guy: he was neither a tree hugger, nor previously very interested in plastic, but he was determined. He now wants to know why things are packaged like they are, and to follow it through to the bitter end.

He took drastic action. Beginning on New Year's Day

2017, he resolved to keep every piece of plastic waste he produced. Anything that contained a polymer was collected and stored – every bottle top, piece of bubble wrap, coffee lid, sliced-bread bag and blunt disposable razor; all the plastic stuff that friends and loved ones left in his home – including his girlfriend's plastic tampon applicators, and crisp bags that friends left in his bin – every single item throughout the year was collected and bagged. If nobody could give him the data on his consumption and recycling, he would get it himself.

One year, and twenty-two black-bin-bags-bulging-with-plastic-waste later, on New Year's Day 2018, Daniel finished his experiment. Then he began the unenviable task of sorting, cataloguing, photographing and analysing the plastic that filled his spare room, and working out what it all meant in the general scheme of things. He called it his Everyday Plastic project, and the fascinating haul is now an art installation and Daniel's photographs have been displayed on billboards at Margate fun park Dreamland; the venue obviously wasn't deterred by their dystopian nature. In fact, they look rather beautiful. Polypropylene bags that previously contained cheddar cheese and polystyrene yoghurt pots take on a certain aesthetic appeal under his artistic direction.

It has to be said that Daniel is only a moderate consumer of plastic. For instance, despite a heavy crisp habit, he was not much of an on-the-go consumer of food and drink, and was already disciplined in the art of carrying and refilling both a water bottle and reusable coffee cup. Yet he still accumulated in the region of 4,500 separate bits of plastic. *That's eighty-six different plastic items a week.* There's a strong likelihood that many of us will top this.

Altogether, Daniel collected 4,490 items made of plastic. Here's a snapshot showing what Daniel discovered about his own consumption:

- Ninety-three per cent (by volume rather than weight) of his haul was packaging for food, drink and other products.
- Sixty per cent of the collection was food packaging (this is almost an exact match for official statistics).
- The non-packaging items included a broken keyboard, disposable razors and some mop heads.

Daniel's accumulated plastic collection includes:

- 639 trays, pots and lids
- 207 bottles/liquid containers
- 284 caps and ring lids
- 450 miscellaneous items. These included a 4 m plastic strip that had been wrapped around a computer.

Some of the individual pieces couldn't be categorised, as Daniel could no longer remember nor identify their original function. This strikes me as a funny consequence of living with your plastic. Such is the breakneck flow of plastic through our lives, we chuck it out in such a hurry that we don't stop to analyse how or why we had it in the first place – we've thrown it away before we even get a chance to remember what its original purpose may have been. One of the fascinating things about Daniel's experiment is that by living with the discarded material, he has been forced to confront all of these questions.

The experiment uncovers other interesting facts applicable to each of our plastic lives, and some anomalies too. Over the course of one year, Daniel collected just fifteen single-use plastic water bottles. But despite the fact that Daniel generally carries a reusable, refillable one – like me, he's a Keepcup fan – fifty-four single-use coffee cups still made it into his inventory. These were the occasions that he forgot his Keepcup.

The Everyday Plastic experiment wasn't conducted just to show Daniel what he consumed. Early on he brought in earth scientist and researcher Julie Schneider to analyse his collection and to explore the lessons that might be extrapolated for all of us, so that we might make different, better decisions in future.

Daniel and Julie calculated the energy used to make all the plastic in the first place at 612.5 kwh. The same amount of energy could power an average fridge freezer for two and a half years, or boil 5,568 litres of water in an average kettle. They also worked out that 207 items (4.6 per cent of the collection) could easily have been consumed in reusable materials or refused altogether. Sometimes we could 'just say no'.

By collecting his plastic rubbish, Daniel has provided a treasure trove of data. He should feel a sense of pride, but I know he's also disappointed by what he found. He assumed, like many of us, that his plastic bottles and tubs would be made into new bottles and tubs. But his experiment showed him that only 1.3 per cent of all the items he collected contained any percentage of recycled plastic.

When he and Julie evaluated the recycling potential of the haul, just 161 out of the 4,490 items could potentially be recycled in the UK: that's just 4 per cent. By their calculations, 289 (6 per cent) would likely be exported for recycling. Meanwhile, 1,277 items (28 per cent) would be destined for old-fashioned landfill, where it would sit in perpetuity. A further 71 items (2 per cent) would go to 'secondary recycling'. But by far the biggest chunk, 2,691 items (60 per cent) would be headed for 'energy recovery'. In other words, they would be incinerated, and a small amount of energy would be recovered for fuel.

Statistically speaking, Daniel and Julie reckoned that fifty of the items, had they been allowed out of the spare room, would have ended up in the sea.[56] I can't get that out of my head. We could each be responsible for fifty plastic items entering the sea this year, unless we change something about the volume of plastics we consume and how we dispose of them. OK, so they cannot be directly linked to

56. To come up with this stat, Julie began with the paper from Jambeck et al, 'Plastic waste inputs from land into ocean' https://www.iswa.org/fileadmin/user_upload/Calendar_2011_03_AMERICANA/Science-2015-Jambeck-768-71__2_.pdf
She says, 'In their model, the way they estimated their famous 8MT a year stat was the following: the percentage of plastic littered in the environment is estimated at 2 per cent of what we consume. So, 2 per cent of your 4,490 = 89.8 items. Then they applied a coefficient to calculate how much of this 2 per cent ends up in the ocean. The coefficient is the percentage of the population of a country living in coastal areas. The hypothesis being that if plastic is littered in a coastal area (say, Margate for instance), it will end up in the sea. In the UK they say that 43,258,889 people lived by the sea in 2016. So, for a UK population of 65,648,100 people in 2016, this represents 66 per cent. 66 per cent of 89.8 = 59.3 items.' This is how Daniel and Julie estimated the 50 number. They stress it's a rough estimation based on the hypotheses listed above.

us, and we did not go down to the beach to lob them in on purpose, but I know I will go to any lengths to make sure that my allocation doesn't go anywhere near the sea from this moment on.

PART 2

6

NEW RULES, NEW TOOLS: HOW YOU CAN REDUCE YOUR PLASTIC FOOTPRINT

It's time to tackle that plastic footprint. We begin with a fond farewell and the presentation of a metaphorical golden carriage clock in recognition of many years of kind service to one of recycling's great icons. Unfortunately this is slightly awkward, as the symbol in question does not know about its retirement. But sometimes we must be cruel to be kind.

The National Museum of American History, Washington, only holds artefacts of profound social, political, cultural, scientific and military history. That is to say, if it wasn't important, it wouldn't be housed there. So I'm heartened to learn that one of the museums houses a collection of more than 1,500 environmentally themed badges, bequeathed by a Mr Gerald H. Meral, a prolific earth-defender (and clearly a badge-wearer). As someone who loves an eco-badge, I could get pretty excited about this collection, but I'll spare you 1,500 detailed descriptions of each badge and instead, let's skip to the main attraction.

The artefact in question carries particular resonance for those of us who consider ourselves pretty sharp recyclers. Against a cream background is set the unmistakable Möbius Loop, the famous arrows, slightly folded at the top, that eternally chase each other, clockwise, round in a circle. Ah, the dear Möbius Loop. Created back in 1970 by Gary Anderson, a student at the University of Southern California, it was rapidly adopted as the internationally recognised universal symbol of recycling. In essence, it indicated that an object could be recycled. As on the iconic badge in the Museum of American History, Anderson's illustration was often paired with the slogan 'Reduce, Reuse, Recycle' (which became known as the 3Rs), and very quickly entered the vernacular during the 1970s.

Together, the Möbius Loop and the 3Rs seemed invincible. They instructed generations in the common-sense ways of the 'waste hierarchy'. After all, what could be clearer than a logical, cascading to-do list in order to avoid waste? In an uncertain world the Möbius Loop and the 3Rs gave certainty. Follow the straightforward three-step guide, choose products that displayed the Möbius Loop, chuck them in the right bin when you're finished with them and you would tread lightly on the planet. They were ever-present reminders and ecological tools, telling us that materials were not something that should be poured into the ground and forgotten about, but that pre-used items should have a plan, a goal for future use.

It is ironic, therefore, that today I'm forced to declare that I consider the Möbius Loop to be in need of retirement.

This has been a long time coming. As the plastic pandemic has unfolded before us, as a society we've come to the stark realisation that *it isn't as easy to recycle as we assumed*. That's not easy for some of us to admit – I for one pride myself on being an excellent recycler. But over the last twelve months I've seen the Möbius Loop displayed on products from single-use coffee cups (when I know that fewer than one in every thousand is actually successfully recycled) to black plastic ready-meal trays that blend into the conveyor belt at the sorting facility, meaning that in most waste management facilities they can't be picked out for recycling without a great deal of effort or the arrival of new technology.

The once-cherished Möbius Loop, telling us that something *could* be, or *has the potential* to be recycled, now seems misleading. But it is liberally applied. In fact, a brand or retailer can slap the Möbius Loop on pretty much anything they like, because in theory any material can be recycled – it just depends how much energy, time and money you're prepared to spend doing it! The question is, *will* it be recycled? Certainly there's a better chance that your waste will be recycled if you sort it, wash it and get it into the right bin. But we just don't know. The moral of the story is that just because your local authority accepts an item, it doesn't mean it's going to be recycled!

In most of the UK – Wales is a glowing exception – we have a Byzantine medley of systems. From one side of Greater Manchester to the other, you might experience an entirely different set of coloured bins, protocols and collection dates. In our plastic special on *The One Show*, in April 2018, our host Matt Baker expressed our viewers' utter bemusement and frustration at the lack of harmony between local authority plastics' collection. He pronounced it 'bonkers'. 'I think we would agree that it is suboptimal,' said Michael Gove, Secretary of State for the Environment, down the line from the House of Commons, 'which is,' he conceded, 'Westminster-speak for "bonkers".' In any case, the well-meaning but simplistic Möbius Loop was never going to be able to shed much light on this carnival of plastic waste practices and processes. In short, there is now too little connection between the venerable symbol and the recycling reality.

In their own inimitable way, industry and retailers have tried to stage an intervention, plugging the gap with . . . yes, more labels! Accepting that the Möbius Loop wasn't getting the job done, they introduced a new sort of on-pack messaging system. This included 'advice', tailored for our patchwork of different recycling regimens. The label is divided into three categories that claim to offer more nuanced guidance to us when we're trying to navigate our local recycling system: there's 'Widely Recycled', for packaging collected by at least 65 per cent of councils; 'Check Local Recycling', for materials collected by between 15 and 65 per cent of councils; and 'Not Currently Recycled', for items collected by fewer than 15 per cent of councils. If you like percentages, and working the odds out while you queue for the till, I guess these are a treat. These are not hobbies of mine, but I do respect the fact that these labels are trying to help and reveal a few more clues. To be fair to supermarkets (and as I don't often give them praise, you may like to note this) the major chains have adopted the new formula labelling with enthusiasm. Unfortunately, large global brands have been reluctant to squeeze these more nuanced labels onto already crowded packaging for the UK market. They say there is simply no room, and want to stick with the Möbius Loop.

Meanwhile, there have even been attempts to reboot the symbol itself and make it more nuanced. In a new spin on the Möbius, when you see light arrows with a dark background (a reversal of the usual) this means that the item is made with recycled material, and that it can be recycled again. But items carrying this symbol are rarer than hen's teeth. The truth is that very few mainstream products are currently made from recycled material. Just to complicate matters further, you may have noticed other flat arrows with a number, usually between one and seven, moulded into the bottom of some plastic tubs, trays and bottles – these just tell you the type of polymer that's in the product, not whether it can be recycled! No wonder we are confused!

In my opinion, the 3Rs are failing us on a daily basis – when we are not failing at them. 'Reduce, Reuse, Recycle' was a bold ideal that carried currency for a long time. But out of the three, 'Recycle' has the strongest voice. It doesn't just cast a shadow across Reduce and Reuse in the hierarchy: it drowns them out. We've become fixated on getting to the recycling, forgetting that we need to stop the superfluous flow of waste in the first place.

This is hardly surprising; every time there is a mention of recycling, we hear about league tables and targets. You'd be forgiven for thinking that the goal was to generate as much plastic waste as humanly possible in order to feed recycling league tables, as if we were in the UEFA Champions League. To add insult to injury, the current system of recycling holds so many pitfalls and pratfalls that we're often set up for failure.

We tried to give the 3Rs and the Möbius Loop a new lease of life. Sometimes you'll see a little 'Tidy Man' popped in the middle of the Möbius Loop. I've even seen an encouraging 'Do the Right Thing' printed alongside it, too. But honestly, research shows this just leaves us even more confused. We should never be contorted with existential angst in front of the local council wheelie bins and recycling crates on the doorstep. It's not fair.

In the past, we've been bamboozled into complacency. I really don't believe that as consumers, householders and individuals we are as reluctant to change our habits as is often suggested. I think we've been blinded by those with a vested interest in the status quo, and persuaded to put our faith in retailers to move to responsible and recyclable packaging: for the people in power to sort out our recycling infrastructure. And been lulled by their promises that it would all work out in the end. It clearly has not.

Plastics have elbowed their way in from every angle, and as if in a frantic game of pass the parcel, we are the ones who are frequently left with the wrapping (but not the prize). 'Plastics is the most complex [of] difficult materials to recycle', according to Douglas Woodring, a global waste expert and the founder of NGO Ocean Recovery Alliance, who also noted, 'most of the world today does not have the ability to recover materials.'[57] For all the assurances and noise that recycling is becoming more innovative and widespread, let's remember that the global recycling rate for plastic is less than 15 per cent.[58] The sorting and reprocessing stages bounce out more bits of plastic along the way. Just five per cent of the material value is what you might call truly 'recycled' and retained for subsequent use.[59] Given what Möbius and the 3Rs have been preaching to us for over forty years, you'd hope it would be a little better. *Fifteen per cent of the globe's plastic waste is recycled, of which just 5 per cent is actually turned into a recycled object or material.* Remember those facts.

Neither the Möbius device nor the 3Rs stand up to muster in this new Age of Plastic. Faced with a tsunami of ever-growing complex polymers and types of plastic packaging, both look hopelessly naive.

I reckon we can do better.

DOING BETTER ...

The following chapters are the practical bit – part manifesto, part tools and techniques to activate your own agency and

57. Renstrom, R.: 'Event highlights plastic recycling issues and opportunities', *Plastics News*, 5 June 2017.

58. Renstrom: 'Event highlights plastics recycling issues'.

59. Ellen MacArthur Foundation, Project MainStream, World Economic Forum, McKinsey & Company: *The New Plastics Economy, Rethinking the Future of Plastics*, 19 January 2016.

play your part in turning back the tide on plastic. Sisters and brothers, we are doing (most of) this for ourselves, by ourselves.

I want to propose a strategy that gets us the best results for every bit of effort that we make.

The speed of plastic production is lightning fast. And this matches the speed of our consumption. In order to turn the tide, we have to slow down the bits of the gargantuan plastic juggernaut that we can control.

Building on the age-old framework of Reduce, Reuse, Recycle, I'm proposing four more practical strategies that we can begin to act on today – conveniently, they also begin with the letter R:

Record collect your own data about your consumption, so that you drive your own success

Replace swap out the ecological hooligans that have colonised your store cupboard, gym bag and commute to work, and swap in the cool high-function, low-impact alternatives

Refuse find your steely inner core to form a robust defence that will stop unnecessary plastic getting into your life

Refill navigate the places and products that bring everyday life up to speed without using disposable plastic

Rethink develop a cutting-edge way of solving plastic challenges. You're part of the solution, not the pollution

Fold in our original three, and I give you our **8R steps:**

Record

Reduce

Replace

Refuse

Reuse

Refill

Rethink

Recycle

Granted, it's a bit harder to remember, but once you focus on delinking your life from plastic, these eight steps will become second nature. Besides, this is all about giving ourselves as many chances as possible to curtail the deadly march of plastic before we even get to Recycle.

7

RECORD

Getting to grips with your household's plastic habit is the first step in turning off the tap of unnecessary plastics that flow into our lives.

Based on the agreed data on average plastic consumption detailed in chapter five, The Footprint of Plastic, page 91, you will have built up a picture of your likely consumption, and should have a fairly good idea of your plastic footprint. Already, you may also have quite a good indication of where you can make immediate changes to cut out some unnecessary plastic in your life.

But before you make any more radical changes, I want you to become your own plastic detective, and commit to this incredibly useful exercise.

GETTING FEEDBACK FROM YOUR BIN

In order to get a full and accurate picture of your household's plastic consumption, you need to gather some data of your own. To do this, I suggest that you start keeping a daily plastic diary. It's not an *actual* diary that you tell your innermost thoughts to – you don't necessarily need to record how you felt when you picked up a six-pack of bottled water, or whether you thought someone looked at you with 'judgy eyes' when you forked out five pence on a carrier bag (though you can if it helps!) – instead, you are going to be keeping it factual and functional.

Simply put, you put aside fifteen minutes each day to register and record as much of the plastic that flows into your life as possible. Keeping your personal household diary for four weeks (or *at the very least* two weeks) will give you an invaluable, most probably shocking, perspective on your personal plastic footprint. If this seems like a big imposition, it's not far off the amount of time you already expend if you take your recycling seriously. When the Proud family in Manchester started their plastic experiment to halve their packaging over two weeks, I noted how much time dad Wayne was spending on their bins: every evening he would meticulously sort through the kitchen bins, picking out the empties that had been put in the wrong waste container.

So while this monitoring and recording process might take a bit of time and a certain amount of effort, trust me: seeing and recording the daily flow of plastic in your household is such an eye-opener, and helps you to really pinpoint problem areas – and score instant, easy wins as well as longer-term, more substantial change. Ultimately it will free you from the tyranny of the recycling bin, saving you precious time and effort in the long run!

Few of us, I suspect, will be prepared to match the commitment shown by Daniel Webb during his mammoth year-long Everyday Plastic project at home in Margate. The information you gather will be a more manageable, succinct version, but the snapshot it provides will still tell you a great deal.

There is something comforting about the familiar noises of the bin lorry and the clinking of the recycling truck, as they go trundling along outside in the street. That comfort, I think, is derived from the fact that somebody is restoring order from chaos. It is all being taken care of on our behalf. Out of sight, out of mind. This recording phase means that we interrupt that process, and confront the reality of how much we consume and chuck away. It can be a tough reckoning, and actually a bit disturbing, to confront that reality. But it is important. The working lives of today's plastics are so short. Your daily diary means that you have a record of what came in and what went out, vital evidence to reflect on long after the bin lorry has been and gone.

The plastic data, based on real-life consumption over the first two to four weeks of recording, will help you make accurate interventions that really work. So remember, after you've recorded the flow of plastic, keep hold of your findings. The secrets of your bin are about to become a precious resource.

INTRODUCING ...
THE PLASTIC DIARY GRID

The diary is in the form of a grid, where you can enter as many pieces of plastic that come into your life as possible (aim for 100 per cent, but you are bound to miss some). My version has eight columns (see example below). Yours might be simpler: to note essentials, the type of object and the number of items. However, the more information you record, the better. In the most time-efficient, basic format,

Plastic / product	Source	Status: **A** = *avoidable* **U** = *useful* **N** = *necessary*	Number of uses: **S** = *single-use* **M** = *multiple-use*
Wrapper: packet pack balsam tissues	*High street chemist*	*U*	*6 individual tissues in pack*
Wrapper: car magazine	*Post*	*A*	*No use (straight to bin)*
Drinks bottle for bio-smoothie	*Health food shop*	*A*	*S*
Plastic toothbrush	*Supermarket*	*N*	*M (as long as battery lasts)*

you might choose just to record numbers of items (assessing the volume of plastics is the most important thing) and then spot-check a few pieces of plastic that have made it into your recycling bin every evening, and record more precise details for five key items.

Draw up or print multiple blank copies of your grid and keep them handy – attach a grid to the fridge, or keep one near the bin to be filled in. (I've included a blank diary grid for you to photocopy in the Further Information section on page 242 at the end of the book, to get you started.)

U = *uninvited* **P** = *purchase (or given with product)* **F** = *free*	**Features**	**End:** **B** = *general bin (landfill or incineration)* **R** = *recycling*	**Number**
P	Thin plastic LDPE. 6 month use-by date	B	12 (Multipack but ALL individually wrapped)
U (no interest in cars)	Thin plastic, heavily inked	B (can't be recycled)	1
P	Thick white plastic	R	1
P	Contains battery	B	1

ON-THE-GO SINGLE-USE PACKAGING

You will also monitor every piece of plastic that enters your life outside of the home – all snacks, drinks bottles, for example, that are bought when you are out and about. Even if you consume it at the bus stop, in the playground, in the lift to your office – every household member needs to find a way of collecting any plastic waste to bring home for recording before it goes into the bin or the recycling. Carry a bag (a reusable plastic one) for this purpose, and bring home empties each night. This means coffee cups, water bottles, wrappers – anything you would normally sling in the bin – bring it home, stare it in the face and jot it down. But let's treat this as an amnesty, so we're not going to judge each other . . . yet.

If you cannot bear to carry your detritus and bring it home, at least make a note of it during the day. Every sneaky takeaway coffee cup, sandwich wrapper and dreaded plastic straw needs to be accounted for, so keep a portable grid on your phone in 'Notes', on your laptop, or in a pocket notebook so that you can add to your tally on the go – on your commute, at work, at school – wherever and whenever plastics try to enter your life. Remember to add any on-the-go data to your master plastic diary at home.

If you live alone, your accounting process should be relatively straightforward. If you're in a busy household, you might be faced with some resistance and even, in some cases, rebellion. Not everybody welcomes change, so prepare yourself. But it's worth persevering. From my experience of working with a number of families in reducing their total plastic consumption, doing the groundwork is worth the effort, not least because returning to the data from the beginning of the project and seeing just how much progress you've made can be a real boost.

SET A GOAL

As you reach the midway point of this period of observation and entering your household's packaging statistics on the grid, you might want to set some goals and a time span for achieving them. At this point, it's also useful to check in and see where your ambition lies.

Are you aiming for the full monty, a plastic-free lifestyle where plastic is all but obliterated? If you're an Instagram user you may have come across members of a zero-waste elite who shun not only plastic, but all unnecessary consumption of materials in order to live a low-impact lifestyle (and sometimes to avoid clutter in their living space). This is now a global movement, and those who are very gifted share their tips for cutting out unnecessary waste with the world on a daily basis. The aesthetic is what I would call 'aspirational', and images often include rows and rows of upscale glass Kimble bottles and jars full of 'essentials' such as tortilla flour and quinoa displayed neatly on white shelving. It is, I promise you, totally mesmerising.

But let's face it, it is not a gritty portrayal of the truth of the fight against avoidable plastic. They do not, for example, look as if they have ever experienced the up-tipping of a refillable coffee cup in their handbag, onto pale skinny jeans, on the number 19 bus. It's easy to be intimidated by (and just a little envious of) their cool, expensive, minimal aesthetic and supremely ordered lives. Yes, it's aspirational, but I have my doubts about how sustainable this type of lifestyle is. It's certainly not a realistic goal for most of us, haring around from one appointment to the next, gathering kids and pets in our wake.

I champion a much more down-to-earth goal, but one not without ambition.

Make a list of the items of plastic that you consider absolutely necessary. Are there any products you are simply not willing to forgo? Perhaps these are blister packs that hold important medication, for which there is no alternative. Or you have a baby in nappies and have already decided you're not going to give up disposables, at least not during the night shift.

After your allocated 'recording' period, tot up your totals for the household. For a couple I would expect you to be running at between 300 and 1,000 items over a four-week period. For a family of four, up to 1,500 items. At a minimum I'd love you cut this in half over the first month.

Whatever your total, whichever the anomalies, let's get started on making a very large dent!

8

REDUCE

Looking at your plastic diary at the end of your recording period should leave you in no doubt: we need to reduce the amount of plastic in our lives. We can't wait for robots to come and scoop it from the ocean and fashion it into trainers (although, intriguingly, this is also happening!). The reduction starts right now, with us.

Stating that we need to reduce plastic packaging certainly won't win me any awards for innovation! On one level, it is crashingly obvious. It's similar to telling someone on a diet to 'just eat less', or someone who smokes to 'just stop'. But these things are all *hard to do*. 'Reduce' needs to be reiterated constantly because we're locked into cycles of daily consumption, and it's so habitual we honestly forget that we are in control and can take our foot off the metaphorical pedal.

The diet analogy is a good one for another reason. Your plastic diary is the equivalent of a food diary – and those have been shown to be pretty effective. There is a connection, it seems, between writing down everything you eat, including the stuff that somehow sneaks in, and taking conscious, everyday, considered steps to stick to a healthy eating plan.

Let's start with the low-hanging fruit. These are very easy wins, and we should have no shame about that.

Begin by ticking every possible box to dent the flow of plastic coming into your home. Increasingly, those boxes are online and on phone apps. Opt out of plastic anywhere and everywhere you can.

SINGLE-USE PLASTICS

Chances are, you're pretty alarmed by the number of disposables you've collected and recorded in your plastic diary. There's certainly a lot of heat at the moment around the idea of creating disposable plastic or alternatives that will make it all better. I have a lot of time for some of these innovations, but we are not there yet. Anywhere, any way you can, you need to reduce your dependency on single-use products made just so that you can simultaneously run for the bus and get your caffeine fix.

A really simple way to do this is by separating the behaviour from the product. Why not make time to eat properly? Next time you grab a coffee, go to a cafe that serves coffee in-house in china cups, sit down, take the weight off your feet and spend ten to fifteen minutes over your drink. It's restorative, plastic-free and just quite a nice thing to do.

CUT DOWN ON FIZZY POP

Across Europe, an incredible 46 billion beverage bottles are consumed every year. Almost half are thought to be plastic single-use bottles, and many of these will contain sugary fizzy pop. Almost half of the 35 million plastic bottles bought in the UK every day are not recycled. If you must indulge, buy a canned drink instead, or do your teeth and the environment a favour and curb that urge entirely! Win-win?[60]

PLASTIC STRAWS SUCK

Single-use plastic straws suck (incidentally, this is the name of a highly successful celebrity-endorsed campaign in the US that has seen straws effectively banned from Seattle). With the exception of those with particular disabilities (who should be offered non-plastic replacements, which we'll look at shortly), most of us do not need straws to drink with, but we have been somehow seduced into embracing the idea that certain tipples *must* be drunk with a straw in order to fulfil a mythological fun quotient. Let's call time on these tubes of needless plastic that are used for an average of twenty minutes, but that represent several lifetimes of trouble for the environment.

60. Sauven, J.: 'If you care so much, Coke, why aren't your bottles 100% recycled?', *The Guardian*, 13 July 2017. https://www.theguardian.com/sustainable-business/2017/jul/13/coca-cola-plastics-pollution-oceans-bottles-packaging-recycling-pr-fizz-greenpeace-john-sauven

Then there are the harder-to-reach bits that we need to try to change.

ONLINE SHOPPING: PLASTIC BAGS AND UNNECESSARY PACKAGING

Assert control over the services you outsource. Increasingly, that includes both our grocery and non-food shopping, as we go online to do our supermarket orders and Amazon famously takes over the world. This remains a tough one to crack. When the click of your order button sends robots off to pick your groceries at fulfilment centres (and will soon include delivery via a driverless van), there are precisely zero opportunities to have a chat with a human to ask if they could kindly lose the excessive packaging and multiple carrier bags. A few online retailers now give the option to select 'no bags'. Others are still relying on recycling carrier bags, insisting that multiple carriers are necessary. In my experience, this is the thing that galls modern-day consumers most.

Another thing that incites 'wrap rage' in even the most mild-mannered consumer are 'over-packaged' deliveries. There are now entire Facebook pages dedicated to Amazon fails, citing nail varnishes that arrive in a cardboard box big enough for a sideboard. These egregious examples of extreme over-packaging appear to defy common sense. There's a reason for that: the retail giant has instead applied commercial sense via a standardised formula to shipping goods. Large boxes are padded out with Air Cushions (i.e., air sealed inside plastic). Air Cushions have quickly displaced newspaper and shredded paper to become a major new plastic peril (although, applying Hobson's Choice, they are marginally preferable to styrofoam, also often seen bobbing about the ocean, with bite marks where fish have taken a toxic mouthful).

In my opinion the online retailers still have a lot of work to do to reduce their plastic footprint, which of course becomes ours. I know I'm one of the last Luddites in town, but this means I buy very little online. I prefer to control the packaging I'm taking responsibility for.

Tick Amazon's 'Frustration Free' packaging option, which is designed to minimise wrap rage at the very least. It will shield you from hard-to-recycle plastics, including the moulded plastic that surrounds many items, such as torches and toys, that can be impossible to get into. In the early noughties British studies revealed that more than 150 people a day accidentally stabbed themselves trying to open packaged products. Moulded, vacuum-sealed plastic packaging film that resisted any attempt to open them without a knife or sharp scissors was very much in the frame. It was calculated that treating packaging injuries cost the NHS a hefty £11 million a year.[61]

61. 'Wrap rage: An open and shut case', *Yorkshire Post*, 4 February 2004.

BUY LESS STUFF

While mainstream retailers and e-tailers remain incapable of providing unpackaged consumer goods, you can of course easily make a dent by simply acquiring less. 'Buy less stuff' is another bleedingly obvious bit of advice, but the moment you don't buy something is always a win for the planet (unless you're buying, say, a solar panel).

DECLUTTER

According to studies, the average American home now contains more than 300,000 items, and I doubt the UK is far behind. There is always an extra small cupboard or shelf that can be put up, and guess what? More plastic paraphernalia will come and nest there.

TAKEAWAY AND FAST-FOOD PACKAGING

According to beach clean data, this is one of the top ocean plastic scourges. Say no to as much takeaway or fast-food plastic coming into your house as possible. Online food delivery service Just Eat, for example, has trialled a pre-ticked box on its app and website allowing customers to opt out of receiving extra single-use plastic items such as cutlery, straws and sauce sachets. Check your preferences on apps you order from frequently and take advantage by ticking opt-outs where possible (it will also remind you of how much apparently mandatory plastic is forced upon us).

A TSUNAMI OF FOOD PACKAGING

Next, we need to tackle packaging. You cannot have failed to notice that most of the plastic that you've recorded in your plastic diary falls into the packaging category. And most of it is wrapped around your food. This presents a dilemma. It seems that to reduce plastic would mean reducing all sorts of foods, and honestly, a person has to eat. So we need to unwrap our grocery shopping from the material that swaddles it.

It's not easy, because Lord, have we developed a thing for packaging in this country! An estimated 2.26 million tonnes of plastic packaging is produced every year in the UK, of which three-fifths (61 per cent) ends up being dumped.[62]

This has everything to do with the fact that we're a supermarket culture. A supermarket economy means a plastic ecology. Supermarkets are responsible for pushing out 800,000 tonnes of plastic packaging a year. Some days I feel like that's all that's going into my trolley, and therefore into my bin.

62. Balch, O.: 'Unilever, P&G and Kraft Heinz criticised for recycling label failures', *The Guardian* Sustainable Business, 8 December 2016.

You're probably worried that I'm going to insist you forswear the multiples, and send you off with a wicker basket to far-flung farmers' markets. This would undoubtedly help – and the more of your shop you can transfer to non-mainstream outlets, the more plastic you can eliminate. I think it is worth trying your local greengrocers and butchers, too, as many are receptive to you using your own containers. But we also have to face facts: many of us are dependent upon supermarkets – not least because they are affordable. Busy lives dictate that you can't always go foraging for less-packaged options. But don't worry, I think you can still cut down. We can be smart about it, using supermarket hacks that mean the plastic isn't dumped on us:

Buy the loose veg!

I'm always intrigued when my friends tell me they don't trust loose fruit and veg because they're worried about hygiene. The bagged, punneted and sealed varieties offer peace of mind. I'm not sure how they know what's happened to it behind the scenes, or maybe they just fear that it has been pawed and squeezed by other shoppers. But I do know this: if there was a genuine risk of serious illness and fatality from unpackaged fruit and veg, supermarkets would not have boxes of it lined up next to the packaged stuff. Let logic win over paranoia. Wash your fruit and veg before use in mild soapy water, or use a special fruit and veg wash, which you'll find in As Nature Intended and other health food stores.

Obviously, not all fruit and veg can be brought to the checkout completely loose. I wouldn't want to queue behind somebody who needed twenty salad potatoes individually weighed and scanned, ditto radishes and tomatoes. But a useful hack to avoid plastic here is to nab one of the paper bags that mushrooms are, for some reason, specially afforded.

Avoid like the plague anything wrapped in thin plastic film. Manufacturers and retailers insist that the shrink-wrap on cucumbers and other closely wrapped produce makes them last longer and is of massive benefit to shoppers. Well, I can't stand it. It makes the cucumber slimy and me less inclined to eat it. The recycling rate for plastic film is just 3 per cent – so I'm pretty sure my council won't take it. I simply won't let it into my shopping trolley.

Keep it simple

Simple foods that don't require refrigeration in the supply chain, or lots of complex messaging, tend also to have less complex packaging (which is difficult or impossible to recycle). Low-fat foods, for example, have been shown to have more complex packaging containing more plastic. This is to do with the fact that they need more artificial flavouring that's kept sealed in through packaging.

Buy in bulk

Buying larger multipacks will reduce packaging waste. Most of the products we buy are repeat products, so where possible buy big – a pack of eighteen toilet rolls instead of four, a single 10 kg sack of rice instead of twenty 500 g bags.

Grow your own

This is not a rerun of the *Good Life* with Barbara and Tom, but if you find, when you consult your grid, that you're accumulating a lot of salad packaging, replacing bagged salad and radishes in punnets with some home-grown produce could be a really easy win. This packaging is a pain. I'm not saying it's not clever: it facilitates long-haul flights and journeys by truck. But that is for the convenience of the supply chain, not for yours. It is really hard to get rid of. I dislike the air-modified packaging that dictates how many salad leaves I get and is crispier than the lettuce within it.

To cut this nonsense out of your life – at least in the summer months – you only need a little space to grow some salad leaves that can be repeatedly harvested, or even a strawberry bed or a wigwam of runner beans. In my own plastic inventory, I found I had a lot of plastic bags for herbs (three or four a week). These are easily grown in pots.

Stop snacking

This might be an unexpected boon, but people who've tried out my strategy so far report that they've also lost weight! Imposing a plastic purge means confronting crisp bags (metallised plastic which can technically be separated but rarely is) and chocolate wrappers. When it comes to the latter, smaller, niche ethical brands such as Divine Chocolate and Seed & Bean have gone out of their way to avoid flexible shiny plastics, replacing these with Forest Stewardship Council-certified paper and compostable versions. The multinational confectionery companies with their incredible size, power and profits, however, continue to wrap the 600,000 tonnes of chocolate treats we get through annually in the UK in wrappers containing polypropylene. There is only one way to tackle this – to swerve it entirely. Nobody is saying it won't be tough!

Cook from scratch, and eat less meat

These are both traditional, eco-friendly ways of eating, but they also help reduce plastic. If our recycling system worked differently, eating ready meals might not be such a great source of plastic waste. But the current mainstream system can't cope with the black plastic trays. Film, as we know, is also a problem. Meat cuts are sold in rigid plastic packaging (only 10 to 15 per cent of all the plastic that is recycled in the UK falls into this sector). The meat industry has also pioneered multilayer film constructions that can get to fifteen layers. In truth, no easily accessible recycling process will have any truck with that. When we talk about reducing, it's not only plastic overall, but also the really complex bits that defy our bins and brains. As I no longer eat meat, I find that I'm spared a lot of the difficult plastic decisions, but that doesn't mean I'm not interested in solutions for meat eaters, too!

Buy from the fish counter and the butchery and bakery counters

And take your own containers. Whereas some might have balked at that, it's becoming much more common. Use an ice cream tub with a piece of kitchen roll for meat, that can be flipped easily into the frying pan.

Morrisons and I haven't always seen eye to eye on plastic – you might remember coconut-gate and the shrink-wrapping argument of 2015 – but they have introduced a scheme where customers can take their own containers into stores. These are weighed at the counter, before your raw meat or fish is placed into them, displacing a load of plastic. At the counter, labels are attached to your tubs, and you take them to the till as usual. Hot on the heels of Morrisons' announcement, Tesco announced it too was trialling a similar scheme. So there are a few reasons to be cheerful.

Finally, consider shifting some of your spend to a store that's leading rather than playing catch-up

As we've established, Iceland – the store, not the country – is already ahead of the game, kicking out plastic from its stores at an impressive rate. By the time I visited the head office just outside Chester and a store at New Brighton in the winter of 2018, the initiative was well under way. Even some of the brand's ready meals were now 85 per cent plastic-free; the film across the top was proving stubborn, but the head of packaging innovation was confident they would soon find a better alternative.

9

REPLACE

I'm not usually the envious type, but I turned green when I came across photographs of Ekoplaza, a small Dutch supermarket in Amsterdam. Instead of the standard walls of technicolour, oil-based, garish plastic bottles, tubs and boxes, was an entire long aisle that was totally plastic-free. Ekoplaza became the first supermarket in the world to consciously create a plastic-free section. It was a sight for sore eyes: 700 products, including milk in glass bottles, sausages wrapped in a compostable plant-based material and loose fruit and veg produce, showed that it could be done. If only we had such an amazing resource at our fingertips here in the UK.

Certainly swapping out plastic products in favour of plastic-free versions is much easier if you have alternatives in shops stocked full of creative solutions nearby. If we don't quite have Ekoplaza, we do have some other good outlets where you can bulk-buy and refill, and more zero-waste stores popping up across the UK. But replacing and switching out plastic products is worth tackling in any case, even if you have to work a bit harder. Why? Because it will do wonders for bringing down your plastic waste. And once you've switched to non-plastic options, I suspect you'll prefer them.

Replace at the right time. Everything that has been made has what is called a 'break-even point', when the amount of resources that went into making it are offset by the number of times it is used. That's one reason why single-use items made from oil-based plastic that expends fossil fuels, make zero sense. It's also why it makes eco-sense to finish the products you already have (the exception being any older cleaning or beauty products you may have hanging around that contain microbeads). Try to reuse, refill or recycle empty bottles or containers.

Overleaf I've listed possible ways to **replace** plastic items. Where possible, I've tried to give a number of alternatives which in some cases **reduce** the amount of plastic in, or on, a product (dematerialisation), or are a switch to a product made of a more easily recyclable plastic.

It is not exhaustive. Given that modern society dictates that we keep a huge number of items, a great deal of which are likely to be plastic, in our homes, that would be ridiculous. Rather, my tips will give you broad-brush guidance on what sorts of substitutes and which product areas need to be tackled, when and why.

For local stockists, recommendations and help at the outset, I recommend that you get online and plug into the plastic-free community movement. Up and down the country they are standing firm, showing the Dunkirk spirit.

This is apt because many of the hottest plastic alternatives back on the table are redolent of post-war frugality. Some are golden oldies that many of us will remember from childhood. Others date back further.

Switch to glass milk bottles

The first glass bottle milk deliveries took place at the end of the nineteenth century. The use of glass for our milk stemmed from the literal desire to show transparency. The customer could see straight away that there was no flotsam or jetsam in their pint-a-day. Now that we take that as a given, there are lots of other things to love eco-wise about the classic glass milk bottle. Once rinsed and returned to the doorstep by the consumer, the bottles are collected as part of the morning milk run, taken to a local bottling plant, sterilised and are ready to be refilled. Each glass milk bottle is reused an average of thirteen times before being recycled.[63]

The win here is in the neat simplicity and the turnaround speed. The plastics industry might argue that PET and HDPE – common types of plastic used for milk containers – can be collected and successfully recycled into new food-grade material. But, as we know, in practice plastic recycling often falls woefully short.

According to Dairy UK, doorstep glass milk bottle deliveries have risen to nearly to one million per day, up from

63. Go to http://www.findmeamilkman.net and type in your postcode to find a milkman near you.

800,000 two years ago. Could this be the biggest comeback since Lazarus?

Invest in a SodaStream

The fight against avoidable plastic tends to centre on bottled water, with good reason. But other carbonated drinks are also a huge source of plastic. Households in Europe with a fizzy drink habit get through between 1,200 and 1,500 bottles and cans in a year. Making the bottles from oil uses in the region of 100 million barrels of oil or twenty times the amount that spilled in the 2010 BP Deepwater Horizon disaster. Home carbonators or soda mixers – SodaStream being the best known – can therefore have a massive impact on your plastic flow. You might remember them from the 1980s, but the next-generation models come in different shapes and sizes: now there's even a portable version, the BubbleCap (from Finland). Switching from carbonated beverages in plastic bottles to a SodaStream is marginally better for your health, too, as the flavoured syrup concentrates use less sugar. And it's cheaper – as a single gas canister will give you 100 half-litre bottles of fizzy pop or plain fizzy water.

Switch to glass containers

For an easy, immediate win, which doesn't need too much thought or planning, swap glass for plastic wherever possible. By 2020 a spectacular 80 per cent of glass will be recycled, and as a material stream it's well on its way. Not only is it easily and infinitely recyclable, it is also proven to be non-toxic as far as food is concerned and very good at preserving.

Glass is not so great for bathroom products, though, not least because of the issue of smashed glass around bare feet. We'll deal with glam and grooming containers separately.

Embrace Tupperware

A good set of Tupperware plastic containers – other brands are available – represents the household workhorse, or pit pony, that will take the strain in your shift away from single-use plastics. It might seem counter-intuitive to invest in yet more plastic, but these are going to offset their creation by working really, really hard and supplanting a lot of 'disposable' tubs and trays.

Deciding which brand to invest in is important, because you will have a view on Bisphenol A (known as BPA), a chemical additive added to plastics #3 and #7 (check the reverse of containers) that has been shown as a hormone disruptor in some laboratory tests on mice. However, other tests have not been able to replicate these results, and that has ended up in a scientific stand-off. Cynics say those tests, which haven't flagged up an issue with BPA, are industry-funded.

The jury is out too on what the implications are for human health. Scientific squabbling apart, you may not want to take the risk. In the US, Tupperware made after 2010 is BPA-free. Here, things are different; the Food Standards Agency is not convinced there is a problem. You may therefore want to seek out certified BPA-free containers – made of plastics #4 and #5. In the UK, Lock & Lock offers an extraordinary range of sizes of BPA-free containers (lakeland.co.uk) and the brand U Konserve (greenpioneer.co.uk) stocks a good range alongside other reusable lunch snack packs, ice packs and tins. If you're suspicious of all plastics (and in some ways I really can't blame you) consider Pyrex and use a tiffin tin for snacks.

Replace as many single-use containers as you can by using your own Tupperware whenever possible. Don't be shy about pushing it under the noses of the people behind the counter at bakeries, meat and fish counters. Ordering takeaways? If you know the outlet well (no judgement here) and collect your takeaway food, tell them you are bringing your own containers along and to let you know when your meal is ready. Remember to label your Tupperware – you don't want it to go walkabout.

Keep your Tupperware in good nick to extend its working life for as long as possible. Hand-wash, preferably, never use in the microwave and if you are dishwashing (Tupperware products are dishwasher-safe), put them on the top shelf, wash seals and containers together so they don't go AWOL. For stubborn staining, scrub them clean using a bit of baking soda and an old toothbrush.

Switch to laundry soap nuts

The last time laundry soap was hot was in the 1920s. Popped into the script of sequential radio dramas by advertisers, it would be liberally referenced by a character (female, of course) and lo! the soap opera was born.

But old-fashioned laundry soap requires a degree of commitment and elbow grease that is frankly terrifying, so realistically it's not a viable substitute to replace the moulded, mixed plastic bottles and tubs with additional dosing balls that most commercial laundry detergent brands rely on.

Instead, try Sapindus soap nuts. These are dried fruit shells containing a natural soap harvested from the Sapindus bush, a shrub related to the lychee, and are very easy to use in place of regular washing liquids or tabs: simply take a small handful of soap nuts, place in a small cotton drawstring bag and shove in the drum of your washing machine. The soap nuts can be bought in some health food stores or from online retailers in a 5 kg sack – that should keep you going! Visit soapnuts.co.uk.

You could also switch to using washing powder from cardboard boxes.

Alternatives to **plastic toothbrushes**

Plastic toothbrushes are difficult – especially the ones that contain a battery and therefore must be removed from the recycling belt (batteries are the biggest cause of fires in recycling depots). Put these in the bin, as we want them to be dealt with properly: lightweight, durable and streamlined, they are almost made for oceans and can travel thousands of miles. Next time, make a better choice: my latest toothbrush is a bamboo version (albeit with nylon bristles) from savesomegreen.co.uk.

Say NO to . . . **microbeads**

About those microbead products. Get rid of them. Now. These are the exception to the wait-and-finish rule. Confine these to their tubes and tubs and bin them (again, these are obviously not recyclable) rather than using them up. We're taking drastic action because we don't want to contribute any more microbeads to an ocean that now contains an estimated five trillion bits of microplastic.

As I write, the age of the mass use of microbeads as a cheap filler, spearheaded by multinational cosmetics companies (at its peak, every year 680 tonnes of microbeads were used in cosmetic products for the UK market) is thankfully coming to an end. Too small to be captured by municipal sewage treatment works, billions have washed into watercourses and found their way into the world's oceans. The manufacture of rinse-off products containing microbeads was banned by UK legislation in January 2018, but retailers were only obliged to stop selling them by the end of July 2018. The best-before dates mean that many of these products will live on long after the on-sale ban, so check your bathroom cabinet and any holiday cosmetics.

You're looking for rinse-off formulas: particularly skin exfoliators, but check for shower gels, toothpastes and moisturisers that may all contain microbeads too. Telltale ingredients to look out for include: polyethylene, polypropylene, polyethylene terephthalate, polymethyl methacrylate, polylactic acid and nylon. These are the most common forms of synthetic microbeads. Be alert for pictures of bubbles on the packaging, and boasts of ultra-deep cleansing: code for 'contains microbeads'. If you're unsure, check older products listed in the UK section of the database beatthemicrobead.org. And if you're reading this on holiday and have bought products while abroad, check the relevant section on the database, bearing in mind that your host country may not have microbead legislation.

Replace your products with reformulated microbead-free versions or ones that contain biodegradable ingredients. There are lots of natural substitutes for microbeads including jojoba, pumice, salt and coffee grinds, and so many brilliant eco-friendly products (some brands never felt the need to introduce microplastics) that serve to highlight how tragic it is that the pollution was introduced in the first place. Unfortunately this is not where the story ends. Products categorised as 'leave-on', including sun creams, lipsticks and shimmering moisturisers (watch out for those containing glitter) are not targeted by the legislation and may still contain microplastics. If you're a consumer of cosmetics, always check the ingredients. Better still, switch your allegiance to an ethical or eco brand that would never, and has never, used microplastics in their products.

Switch to soap bars

The giant personal care brands that I'm afraid must be held responsible for filling our oceans with microbeads have also encouraged us to migrate from good old-fashioned bars of soap towards more expensive, bottled liquid soap and gels which are, surprise, surprise, abundantly packaged in oil-based plastic bottles. Many of these do not make it into recycling because they are made from strange and difficult plastics, and some contain extra pumps and complicated dispensers.

Move in the opposite direction and go back to bars of soap. For extra brownie points, buy an eco-friendly soap wrapped in paper rather than a plasticised wrapper. If you hate the ectoplasm that soap leaves on the side of the sink, buy a ceramic or natural resin soap dish.

Go naked*!*

Follow the rather startling command from cosmetics brand Lush. From toothpaste in jars, conditioner and shampoo in bars rather than bottles and a cheeky soap produced in the shape of a plastic soap dispenser, Lush has pioneered plastic-free, or 'naked' in Lush parlance. One shampoo bar provides up to a hundred hair washes, so that displaces the equivalent of three 200 ml bottles of standard liquid shampoo.[64] The brand is loved by kids and young adults, and has the added benefit of being low-cost and easy to track down – you can smell the stores a mile off. This is not because they are fans of olfactory marketing, pumping out perfumes into the outdoor environment, but simply because their ebulliently scented products are unwrapped. Transported to your own home, this also saves on air freshener!

Switch to cloth nappies

Eight million disposable nappies are shovelled into landfill in the UK daily, and – regardless of what it might claim on the plastic packaging, and some do make claims of biodegradability – here they will fester for hundreds of years.

64. UNEP (2014) 'Valuing Plastics: The Business Case for Measuring, Managing and Disclosing Plastic Use in the Consumer Goods Industry'.

Waste materials need oxygen to biodegrade, and there's not much of that in landfill.

Cloth nappies conjure up images of babies' bottoms swathed in white terry-cotton squares secured by oversized safety pins like in *Call the Midwife*. But today's reusable nappies offer a number of different and much more practical, washable and reusable options, made from varying naturally absorbent materials such as bamboo and hemp. Some 2.7 kg of raw materials are used in a full-time set of reusable nappies, as opposed to 120 kg if a child is in disposables.

The initial financial outlay might feel steep, but over the long term the investment compares favourably to the cost of disposable nappies. If you balk at the upfront costs of investing in two to three years' supply of nappies, there are also local 'nappy library' schemes which offer budget-friendly options. To help evaluate reusable nappy options, go to Go Real (goreal.org.uk).

Switch to cardboard toys

As a doting and easily manipulated aunty, I am forever being manoeuvred into toy shops and down the toy aisle in supermarkets by my nieces and nephew, who talk me into buying them plastic tat as a matter of urgency. I feel guilty. Not only do I know their parents hate yet more plastic coming into the house, but also I really don't want another generation to get hooked on plastic.

I wonder, then, if Nintendo, surprisingly, has answered all our prayers. I had high hopes for this year and our mission to turn the tide on plastic, but I hadn't reckoned on Nintendo making it quite this cool. Labo, the brand's range of cardboard toys, is a game-changer in every sense. It features a range of pop-out sheets of colour-coded templates, called Toy-Cons, made from regular cardboard as opposed to plastic. With rubber bands, string and a lot of patience, these are constructed into all sorts of amazing toys, including a remote-controlled car, a giant bug, a small piano and a fishing rod. Then they're brought to life by slotting in a games console, the Nintendo Switch. It is being touted as the toy that will revolutionise the video game market, and it seems to have single-handedly dealt a major death blow to plastic. Wow.

10

REFUSE

In a culture that takes a minimum of twenty-nine 'pleases' and 'thank yous' for the most basic of transactions, refusing anything, let alone plastic, can be unexpectedly hard. It took me a little while to master the art of the casual, breezy rejection. I used to purse my lips and give too much information, as if I was going to deliver a sermon, adding, 'You know that paper is actually coated in BPA, a form of plastic additive that has been highlighted as a possible endocrine disruptor?' Activism Tourette's. Beware. Do I want a till receipt? 'No thanks, I'm fine.' By the way, it's completely true: a 2014 study found that almost all receipts are coated in our old friend the plastic additive BPA.[65]

When it comes to refusing plastics, we have grown in confidence and stature thanks to the five-pence levy on

65. Horman, A. et al.: 'Holding Thermal Receipt Paper and Eating Food after Using Hand Sanitizer Results in High Serum Bioactive and Urine Total Levels of Bisphenol A (BPA)', *PLOS* 22 October 2014.

plastic bags that was introduced in England in October 2015 (Northern Ireland, Scotland and Wales introduced it long before). Immediately before its introduction, people fretted about how they would say 'no' to the habitual free plastic bag, as if it was an epoch-ending scenario. In actual fact it was fine, dare I say it; relatively painless even. Happily we now get the concept of refusing plastic.

DESIGNED FOR LANDFILL

Some plastic products are just so badly designed that I've taken a stand and simply refuse to have them in my life. Indeed, as well as refusing the plastic bags, wrapping, receipts or single-use cutlery that we're given as a matter of course, I want us to say no to bad design. Given that we are such experienced and demanding consumers, it is a bit weird that we are so accepting of products that are essentially designed for landfill. We are forever inviting ghastly articles into our homes, where they take up precious space and do not earn their keep.

Bad products made from plastic are particularly pernicious, wasting our money, our space, our time, our resources and – to add insult to injury – winding up as a piece of junk destined only for landfill. Let's not forget that these things were actually *designed*, and that that discipline has rules and protocols and centuries of human ingenuity. According to Dutch academics in design theory, 'Design is the activity to convert a Mission Need Statement or a set of User

Requirements into a product, which meets the stated needs or requirements.'[66] The Dutch know a thing or two about design and engineering, so I trust this definition. However, I would love to see the mission statement for the stuff that makes it on to my list of shame, and yes, I actually have one. Not only did I find writing it strangely cathartic, but it also made me decide to put a bit more energy into getting a global ban on the items on my list. Let's be clear: my goal is to eradicate them. Here is my top three:

The airplane coffee cup

There's the good, the bad, the ugly and the downright ridiculous. 'Have you had a coffee with us before?' asks the stewardess on the BA017 short-hop service to Zurich. I assume this is a loyalty card scheme. But no, she actually needs to give me an induction into how to drink from the airline's single-use coffee cup. This appears to be a large hunk of Polymer #6, resembling a toddler's sippy-cup with a large sculpted lid. The invention is 'patent pending' – according to the many instructions and flaps on the lid, and I can see why they might want to think about it.

The stewardess shows me where to drink from, and from where not to drink. If I remove the lid I will be in extraordinary trouble, scalding myself with molten-hot liquid: I am also warned that all the coffee bits will fill my mouth and it will be unpleasant. Ditto if I drink from the wrong flap. The designated drinking hole is covered in a mesh, but the whole

66. Hamann, R.: 'The Function and the Design Process', 15 International Cost Engineering Congress, Rotterdam, 20–22 April 1998.

operation is completely counter-intuitive. I tentatively swig from the mesh-covered portion. My mouth is instantly filled with grainy bits of coffee in any case. Is this the worst use of plastic ever? I think I've found a new plastic nemesis. Spork, you're off the hook.

The pouch

The stand-up pouch with valve is considered a breakthrough plastic innovation in the industry, and it's one of the worst things I've ever come across. Containing such delights as pet food (I don't see why this can't come in a tin), the pouch is over-engineered. It boasts a 'bottom gusset' affording 'self-standing strength, degassing valve and excellent visibility through a custom-cut window'.[67] Wow. I still think a tin did the job pretty well, and my dog isn't much impressed. As any material is theoretically recyclable, I can't say that these pouches are impossible to recycle. What I can say is that they are challenging to recycle in most local authority schemes. In the US market they are sometimes promoted as 'landfill friendly' because they flatten down and take up little room.[68] This does not sit well with my more ambitious vision of us as plastic reducers.

Disposable plastic toothbrush with battery

Dental lore says that we must replace our toothbrushes every three months to avoid gum disease (by the look of those scary adverts this is very much worth avoiding), so I

67. The Bag Broker, https://www.thebagbroker.co.uk/stand-up-pouch.html
68. https://www.standuppouches.net/blog/can-kraft-stand-up-pouches-be-recycled

see the imperative to chop and change toothbrushes. But these? They're an outrage. The battery, moulded into the handle of the toothbrush, is non-replaceable. This means the product calls time on its usage, not you! (I'm sorry: I think that's rude.) According to several design experts I consulted, the position of the battery and the moulded plastic design of the whole brush means that the vibration doesn't aid teeth-cleaning: it just makes the handle vibrate.

When it has run out of battery, this toothbrush must be thrown away, and because the encased battery cannot be dealt with at normal household recycling centres for safety reasons (ditto energy recovery facilities), the best-case scenario for this toothbrush is landfill. But a lot of people will assume it can be recycled. Confusion around the bin leads to fugitive plastic and, alas, I see a lot of these on beach cleans. If you were designing something that was relatively light, durable and could travel long distances as ocean pollution, I'm afraid these battery-operated plastic toothbrushes would fit the bill perfectly.

HOW TO BECOME A PLASTIC REFUSENIK

THE GREAT BRITISH UNWRAP

Sometimes you may still feel that nobody is listening, and want to make a fuss. This is where more dramatic action is required. My grandad was on the money here, with his small-scale one-man protests against packaging, unwrapping anything he thought excessive at the cash till. Actually, he was following a hearty tradition started in the 1980s by a group of Austrian housewives who started a wave of protests by unwrapping all their shopping in stores in a coordinated action. It rattled Austrian retailers so much that they began to get rid of plastic nearly forty years ago (with limited success; there too it has crept back in). This form of protest, though, the staged coordinated unwrap, has made a comeback. This year has seen spirited protests in Bath, Frome and Chard in Somerset, and in Bideford in North Devon (what is it about the South West?).

Thanks to social media, these public acts of protest have had a lot of traction. There are some rules: organised, coordinated plastic protests are best (check out local action groups on Facebook); all goods must be paid for and packaging should be gathered up. The so-called 'plastic attack' in Bath netted three entire trolleys-full of plastic, reinforcing the point that there is an alarming excess. Today's unwrappers tend to go one step further than the original Austrian *Hausfrau* activists. Protesters tend to bring their own reusable containers and transfer products into them. Milk is poured into glass bottles, cheese liberated from plastic and put into greaseproof paper. I like the way this moves the conversation on, as it then becomes about alternatives and a different way of shopping, and is not purely supermarket-bashing. Watch out, though. A group of unwrappers in Ireland was recently threatened by Retail Ireland, the organisation representing major retailers that posited they were in danger of flouting litter laws and could be liable to prosecution. We can only hope that they choose to spend their time figuring out ways to use less plastic, rather than prosecuting those who take a stand.

I urge you to write your own plastic list of shame. These are the products you wish to excise from your life. Pin it to the fridge or cupboard, and use it as a statement of intent. No more will these get past your threshold!

WRITE A LETTER

I don't live in Tunbridge Wells, but I am so disgusted by the constant examples of over-packaging I see every day that I've taken to writing letters to brands. Whenever you see packaging that is totally over the top, take up the pen and write to the company responsible. It needn't be a formal letter – a tweet will often be just as if not more effective. Add the hashtag, #reducepackaging to your tweet, and make sure you copy in your local council and trading standards office (if they are on social media). I explain why this is important below.

Of course, you have no guarantee that your carefully worded tweet won't be ignored by the brand or organisation in question, or your letter shelved in the famous File Thirteen (aka the bin). But persistence is key, and many brands are acutely alert to criticism of the amount of plastic they use in their packaging.

If you have time, research the company's position on plastic. Increasingly, many corporates have a plastic reduction policy. In your letter, email or tweet, point out how the offending plastic goes against their policy and/or ethos. I was recently prevented from entering the flagship Apple store in central London until I had put my umbrella in a single-use plastic 'umbrella bag'. Nor was I allowed to leave my umbrella by the door. The umbrella cover was mandatory if I wanted to come inside. I later emailed the store manager pointing out that, given the work they had done to remove plastic from the packaging used for Apple products, this was ludicrous. He agreed, and is now asking for a policy change on leaving umbrellas by the door.

REFUSE UNSOLICITED DIRECT MAIL

Neither am I above returning annoying, egregious packaging to the sender. Much of it arrives uninvited by direct marketing anyway. Register with the Mail Preference Service (mpsonline.org.uk). The direct marketing industry is obligated to pay for this service and ensure your home does not receive unsolicited mail. As, increasingly, that mail is either wrapped in plastic or printed on plasticised paper that cannot easily be recycled, this is an important way of stemming the flow of plastic through your letterbox.

Here's a sample letter that I wrote to BMW (I am still waiting for a response):

Dear Paul
You recently sent a copy of the latest edition of your magazine to me. I'm not in the market for a new car, so it was slightly optimistic, but nevertheless thank you.

However, the packaging caused me huge concern. The use of plastics is completely unsustainable. Lightweight plastic films and wraps remain one of the most pressing parts of the issue.

There are two problems here: the wrap is printed and displays a lot of info. It has an extra picture of a car which is already on the magazine cover, so is needless. It also shows information directing me to a change in the data protection laws reminding me to 'opt in'. But there is no room on this pack for information about recycling the plastic you've dumped on me.

As the wrap is printed, this causes extra problems. Plastics containing metal, oily food residue or inks, that add up to five per cent of the weight, are not suitable for recycling. So you've produced an avoidable and very annoying piece of plastic, which I am returning to you, enclosed.

It also goes against your brand values, which rest on clean, green, intelligent design.

I'd be really keen to hear your targets for plastics in marketing communications, and if you don't have any, I would be delighted to suggest some.

Very best,
Lucy

THE LETTER OF THE LAW

Don't be afraid to be much tougher. The law is on your side, with legislation originally implemented over twenty years ago to combat a surge in packaging, particularly in plastic. The Packaging (Essential Requirements) Regulations (PERR) might not be very widely known about, but they are supposed to prevent over-packaged products.

The relevant clause states: 'Packaging volume and weight must be the minimum amount to maintain the necessary levels of safety, hygiene and acceptance for the packed product and consumer.' From shrink-wrapped coconuts on a plastic stand, to tiny purchases in huge boxes padded with air cushions, between us I'm certain we have plenty of examples of packaging that risks flouting this law.

The offenders ought to be reported to your local Trading Standards Officer, who is responsible for taking action. From time to time, local councils even call for examples of bad packaging. I last saw this a couple of years ago when Birmingham Council asked people in the region to send in pictures and details of over-packaged products via social media. But why wait? Jo Swinson, MP introduced the Packaging Reduction Bill to the House of Commons in 2007 (unfortunately the government didn't back it), but she has continued to campaign against over-packaging. You can download a template letter to Trading Standards from Jo's website http://www.joswinson.org.uk/excess_packaging.

Given the heat and energy around avoidable nuisance plastics, now is the time to take this on again. In April 2018, Defra doubled the maximum fine[69] for those who chuck plastic waste from their cars to £150. Isn't it time the law was beefed up to take on those who produce unnecessary waste in the first place?

69. House of Commons Environmental Audit Committee: 'Disposable Packaging Coffee Cups', Second Report of Session 2017–19, 19 December 2017, HC267.

11

REUSE

The polar opposite of the detested single-use plastic item is a reusable one. In this disposable world, reusing an item multiple times is still considered an act of defiance. But a culture of reuse is a culture of purpose, and gets a big thumbs up from me.

Reusing diverts plastic, and takes pressure off the recycling services. Having witnessed recycling centre staff bracing themselves for another influx of empties of myriad polymers in all shapes and sizes, some of which are difficult to sell on the world market, there's something to be said for cutting the system some slack. Similarly, while energy recovery or incineration plants are painted as avaricious monsters that constantly need feeding, there are limits to the amount of plastic they can burn. The system is clogged: help it out.

As we know only too well, perhaps, a lot of that plastic is virtually indestructible. Wherever possible, use this fact to your advantage.

When you do buy plastic, assess whether there's any reuse value in the packaging. I don't, for example, fancy your chances of reusing a sandwich wrapper. But a nice sturdy tub, well, that's a different matter. If it has a good seal and is of a more rigid plastic you can use it for leftovers for the fridge or freezer, at meat and fish counters or for packed lunches instead of Tupperware.

In reusing plastic products we are taking responsibility for our plastic footprint. It's one of the areas of our plastic life where we can use our own judgement, and just get on with it without outside interference. So, altogether now: 'rinse and repeat' as many times as you can.

REUSE CHECKLIST

1 Clear plastic products are the obvious place to start to reuse from home. Once thoroughly washed and rinsed, you can easily see when the container is empty and clean. A quick check on the bottom of the bottle will confirm that it's made from PET, the right type of plastic (this will be displayed in the triangle that says PET 1). Given that most water bottles are made from PET, and 38.5 million plastic bottles are used every day – just over half of which make it to recycling, while more than 16 million are put into landfill, or become fugitives – we have an abundance. Everyone should invest in a durable reusable water bottle, but if you've left it at home you can offset the indignity of buying bottled water by reusing the bottle ten times . . . minimum. You can even take your refillables through airport security. This includes empty standard water bottles that you want to reuse. Empty and carry for a safe passage. Write your name on the bottle, so it doesn't get thoughtlessly chucked in the trash, and prepare to fill from the tap or a refill station.

2 Don't be squeamish about reusing plastic, especially water bottles. We need to put to bed the frequent rumours, especially on social media, that to reuse water bottles is to flirt with a toxic force field. These alerts warn that dangerous chemicals will leak from the plastic into the drink, especially, for example, if a bottle is left over a significant period in direct sunlight. According to the chemists I've consulted, this is bunkum. The pernicious chemicals in question will generally emit a fishy odour, so at least your nose will be heavily alerted if you do come into contact with them. Furthermore, the toxic chemicals in question are not in found in PET (the type of plastic used in 99.9 per cent of bottled water). More typically they are being found in PVC, which we would never want to use in a water bottle. Of course, leaving a bottle in sunlight could produce a bit of a bacterial flourish after a few days, but common sense and normal hygiene prevails – if you are concerned, sterilise water bottles in cold sterilising solutions before reusing.

3 Never assume that containers are dishwasher-safe. Remember that these have been put on the market with the intention that they are disposable, so the quality of some plastic may not survive high-temperature dishwasher cycles. Some sturdier plastic containers can be placed in the dishwasher on the top shelf on the low temperature setting, but never stack any plastics at the bottom near the element, where they risk melting. It's much safer simply to soak and wash in a bowl of soapy water, rinse well and leave to air-dry before reuse.

4 The plastic ready-meal tubs, trays and pots are made of lower-quality plastics so aren't worth recycling as they don't have much resale value on the global market. Try to reuse them instead, in ways that displace the need to buy other, new plastic products. Be creative. For example, put old plastic tubs to use as paint pots. Clean, ready-meal trays with drainage holes carefully poked in the bottom are ideal planters for seedlings.

5 Trigger sprays grace the top of many cleaning products. Respect, salvage and cherish them. They took a lot of plastic to make and are rather elegant bits of engineering; most contain a spring, hence the trigger action, and are way too useful to last the lifespan of just one bottle before being discarded. As long as the cleaning product doesn't contain any potentially dangerous or toxic ingredients, thoroughly wash out the spray bottle and refill it with your own simple surface or window cleaner made from white wine vinegar.

6 Take your own reusable cutlery with you on the go. Reuse straws, single-use cutlery, stirrers and carry a Spork (if you must). Air and rail travel generally is fraught with single-use plastic. Keep a portable cutlery set in a pouch in your bag, to protect them from handbag lint. These are simple to sew yourself. I was lucky enough to be given a ready-made one by Bettina Maidment of Plastic Free Hackney. Not only is it covetable – made from a sort of embroidered toile de Jouy fabric – it's a game-changer.

Every time I reuse my plastic cutlery (originally liberated from Pret à Manger) I feel like I'm winning.

7 Some plastic products are made to last, and have a lot of charm. Many have been retired far too early and dumped. Plastic toys, kite-marked, from known brands and no more than a decade old should surely be given a stay of execution! If they begin to look a little grubby, most of the more robust plastic toys – Fisher-Price Play People and Weebles, for example – can be popped in the dishwasher on a medium temperature cycle and they will come out like new. Again, use cold sterilising solutions if you want to be sure.

When you think of all the Lego sets, play farms and Barbies and Sindys that populate the nation's lofts, it is tempting to get them all out and find them a home, or at least donate them to charity shops. A word of caution, though, before you reuse vintage toys: legislation on plastic use in toys has changed over time, and rules governing the types of plastic imported to the UK are now more carefully controlled. When it comes to toys that are going to be heavily handled and perhaps put in mouths (I wouldn't reuse any teething toys) exercise caution. Vintage plastic dolls in particular should be handled with care. If you have an elderly doll in your house (dating from the early 1960s) and you notice he or she has a shiny face, they could be suffering from Greasy Doll Face Syndrome. This phenomenon indicates that the polymer and oil have (beneath the surface) started to part ways as the polymer slowly degrades. These dolls, sadly, are ready for the toy box in the sky.

8 It's not until you start keeping a watch on plastics that you notice which mass-market brands are actually going backwards. The biggest players in feminine care have recently switched from cardboard tampon applicators to plastic ones in pastel shades. It's as if they were hell-bent on bringing yet more plastic into our lives. They've certainly had an effect on our coastlines. In their 2016 beach clean-up, the Marine Conservation Society found twenty tampon applicators and sanitary items per 100 m of shoreline.

Make a stand by switching to reusable menstruation products. The move to a more eco-friendly alternative to tampons has been spearheaded by Mooncup, a reusable menstrual cup that has gained a serious fanbase.

12

REFILL

Everybody, I think, gets the logic of refilling – it's pretty straightforward. But sometimes we're tripped up by very strange concerns. A report commissioned by Brita UK (the water filter people – see below) and Keep Britain Tidy revealed that almost a third of millennials (29 per cent) said they don't use refillable water bottles because they find them 'too heavy'. Puh-lease. I think we can bear the weight! Besides, refilling, replenishing, decanting – however you prefer to describe it – is a game-changer. Give it a chance and it'll soon be embedded in your daily routine.

The refillable culture was once completely instinctive. Only a generation ago, mandatory deposit schemes for bottles and refillable containers were commonplace. In fact, the system was set up to work this way. But soft drinks manufacturers in particular didn't like this sustainable status quo, and by 1979 Coca-Cola and Pepsi had between 1.5 billion and 1.7 billion plastic soft drinks bottles on the US market, and were making inroads into other territories.[70] The records show that they had to lobby quite hard for equal shelf space with refillables (namely glass), raising challenges on competition and unfair trade grounds. It worked: acts were amended, legislation changed and plastic bottles began to flood into our lives. As consumers, we took to them – too well, in fact.

Refill culture never quite died out in the way, I suspect, many multinationals wish it had. If you've lived in one of the determinedly green constituencies – Hay-on-Wye; Totnes; Brighton; Hebden Bridge (I've lived in two out of the four) – someone you know will possibly have had the same laundry detergent bottle for fifteen years, and every few months will assiduously fill it up at their wholefood store's refill station. And all power to them.

We need to join them, because when it comes to refills there's a sense of use it or lose it. Anita Roddick, the irreplaceable force in UK ethical business and founder of The Body Shop, used to claim that, in the early days, she offered the store's shampoo and conditioner refill service because she didn't have money to buy extra bottles (this is a useful point, by the way, if you are contemplating a beauty start-up!). Even so, as the business grew, the refill model remained.

In 2002, however, The Body Shop stopped refilling due to lack of interest: only one per cent of customers were using the service. But major high-street brand MAC Cosmetics

70. Turner, J.: 'Plastic pop bottle decision anxiously awaited by firms', *Globe and Mail* (Canada), 17 August 1979.

runs a 'Back to Mac' scheme, offering a free lipstick for every six empty lipstick containers you return to a MAC store or send in to MAC online.[71] Today we need to support companies that offer refills, and encourage new ones by showing them that this is what the market demands. In short, embrace refillables as enthusiastically as you can.

First, let's tackle the two major causes of concern and eliminate them from your plastic diary: single-use takeaway coffee cups and plastic water bottles.

BUY A REFILLABLE COFFEE CUP

Coffee culture has gone crazy in the UK over the last twenty years. By 2025 the number of coffee shops is forecast to increase from 20,000 to 30,000,[72] and unless we do something, that means more cups. This is horrifying. Already the total amount of annual coffee cup waste in the UK is enough to fill the Albert Hall.[73]

Many of us assumed that, because they look paper-y, coffee cups could just be plonked in recycling bins along with newspapers. Not so fast. The fact that the 2.5 billion coffee cups produced in the UK every year are not easily recyclable came as a shock. But once I'd seen single-use coffee cups being made, the reason why was less mysterious. To make them leak-proof and heat-resistant, plastic is poured onto

71. https://www.maccosmetics.com/giving_back/printable_form.tmpl
72. House of Commons Environmental Audit committee, 'Disposable Packaging: Coffee Cups', Second Report of Session 2017–19; 19 December 2017, HC 657.
73. HOC Environmental Audit committee, 'Disposable Packaging: Coffee Cups'.

paperboard and the materials are fused together. The cups are then punched out of the laminated cardboard by machines. It's a rapid-fire production process. The cup is in use for the time it takes you to drink your beverage and then they're binned, not very successfully. Only two paper mills in the UK have the technology to separate the plastic from the board. I suspect they're rather busy. Only one in 400 disposable coffee cups are recycled.[74]

Takeaway coffee cups come with an added complication: the lids are rarely taken into consideration, but they are single-use plastic and even less likely to be recycled than the cup itself. That's saying something.

Opt out of this craziness as soon as you can by investing in a refillable. Sometimes this can seem like a considerable investment as some tend to be pricey, but you could consider it a saving, as you can make a tea or coffee at home and take it with you on your commute. Cabin crews are not receptive to making you a brew in your own cup. Understandably.

I favour a Keepcup. This Australian brand made from polypropylene earned its stripes on university campuses and has become one of the breakout designs of the refill movement. There are a couple of things I like about Keepcups. Baristas like them because they're designed to fit under a coffee machine. Sounds obvious, but I don't want to be the one holding up the queue in the morning, glowered at by an angry barista. It also means that they're not tempted to use a disposable cup to make the coffee, pouring it into

74. House of Commons Environmental Audit Committee, 'Disposable Packaging: Coffee Cups', Second Report of Session 2017–19, HC 657. Published on 5 January 2018.

your refillable (which completely defeats the object). The design is also really well thought through, and to be honest, I think it's the coolest, but check out other similar makes and models, especially e-coffee cups made from bamboo rather than plastic.

BUY A REFILLABLE WATER BOTTLE

As far as I'm concerned, the need to get away from single-use bottles of water can't be overstated. UK households get through 13 billion plastic bottles a year, including all beverage bottles and toiletries. But of that total, 7.7 billion are water bottles. Your plastic diary is likely to record three bottles for every adult per week (based on averages). Even if it's less than that, get yourself a refillable water bottle. We're on the cusp of a water revolution that will hopefully spell curtains for billions of single-use bottles.

I've road-tested a few refillable water bottles. I was looking for one that didn't tip down my front when I absent-mindedly took a drink, or leak in my bag (I've experienced both). There are some covetable, highly polished, expensive-looking ones around but I'd only worry about scratching and scuffing. My favourite is from Klean Kanteen, as it's robust and very reliable and doesn't leak. When I go running I switch to a lightweight plastic refillable that fits around my hand.

There's a lot of research that shows that more people would refill from the tap if their refillable bottle had a filter.

To which I reply, what are you waiting for? There's plenty on the market that tick this box. Bobble is a great on-the-go water bottle complete with a filter.

WATER REFILL STATIONS

Once you've got your bottle, where do you refill it when you're out and about and thirsty? In 2015 a handful of cafes in Bristol started offering tap water refills to anybody carrying a refillable bottle. By the spring of 2018 this had ballooned to 1,600 different venues, stretching from Dumfries and Galloway to the Isles of Scilly, all displaying a blue sign signalling that they'd be happy to fill up your bottle. These include Costa Coffee and Premier Inn (part of the same group). Refill.org.uk will point you in the direction of refill participants near you, or anywhere you're planning to travel to. In practice, whether or not they're part of the scheme (and why wouldn't they be?) most restaurants and cafes are happy to fill up a bottle with fresh drinking water for you. And it's not only the hospitality industry that is cottoning on: the first high street chain to offer itself as a water station was Neal's Yard, the organic cosmetics chain.

WATER FOUNTAINS

If you're shy about asking at a cafe, soon you'll be able to fill up at public water fountains as increasing numbers of local authorities commit to putting them in place. Drinking fountains have a long and venerable history. In 1859, in response to an outbreak of cholera, well-heeled London residents formed the Metropolitan Drinking Fountain and Cattle Trough Association, and private money paid for the first drinking fountain. By the end of the nineteenth century there were 800 drinking fountains in London. Lucky them (apart from the cholera).

I'm really glad drinking fountains are coming back into vogue – theoretically at least, as they are yet to be installed. Among others, Michael Gove and Sadiq Khan have signalled their commitment to installing a new generation of water fountains across London in the fight against single-use plastic water bottles. As I write this, in late April 2018, Reading Council has just announced a collaboration with Thames Water to provide a network of drinking fountains. When we do eventually get them, don't necessarily expect to see the old Victorian repro style, complete with cherubs and curlicues: next-generation water fountains are hands-free bottle-filling stations, ergonomically fashioned to fit reusable water containers, complete with automatic sensors that stop the flow of water when the bottle is full. Can't wait!

POSH WATER

In my own home I keep it low-tech and just use the tap, although I have recently bought a copper water jug. I'm a big fan of UK tap water – it has the highest treatment standards in the world, and they are higher than bottled water. But I'm always surprised at the vehemence with which some of my friends take against it. There is some research that suggests the type of tap water you like is derived from early geography. So if you grew up in Derbyshire, for example, you're predisposed to a peaty taste. For those repelled by tap water, you can always refill from a filter jug like a Brita. Or you can go really high end. The Ovopur water filter comes in at a cool £595, is made from porcelain by Aquaovo and has an AquaCristal filter that lasts up to four months. In the UK it is distributed by Stokes Tea & Coffee.

JUICE IT

Once you've got into the habit of buying loose, unwrapped fruit, if you are big fruit juice consumers in your family, the next step is to acquire a juicer and cut down on fruit juice bottles and other packaging. Fruit juice is predominantly sold in Tetra Pak cartons, and the packaging giant produces 184 billion single-use cartons annually. These may look like paperboard but they contain 20 per cent polyethylene and 5 per cent aluminium, as well as the plastic spout and lid, and I still see a lot of them thrown into paper recycling. To be fair, Tetra Pak has invested heavily in recycling collection points in the UK (for link, see page 206), but they are a complex form of packaging that takes a lot of energy to reprocess.

Much better to start juicing at home and bypass packaged juice in Tetra Pak containers completely.

BULK-BUY AND REFILL SYSTEMS AT HOME

Reusable water bottles and a Keepcup or two are the obvious household essentials. The next step is to set up and buy into other refill systems that can really make a dent in your plastic consumption. Refilling and/or buying in bulk, to decant into reusable containers-for-life at home, is preferable to simply switching to a brand that uses 'biocompostable' plastic in their packaging: you will still have to consign all those smaller containers to recycling, biodegradable or not (see page 227 in chapter 16, Squad Goals, however). By switching to buying in bulk-size quantity and using sturdy, refillable containers-for-life, you will cut down on your plastic consumption considerably more radically.

You need to know what's really practicable within your routine and your space. The potential with refillables is that you get into a habit that you can sustain. Start by clearing a couple of shelves or a cupboard in the kitchen, garage, utility room or garden shed, if you are lucky enough to have any of these, and dedicate it to refill containers. By starting slowly, you'll save yourself from splurging on expensive Kilner jars and refill kits – always a temptation. I'm only a step away from trying to turn myself into a zero-waste Instagram princess!

Buying in bulk and/or refills is ideal for household cleaning and laundry products, and also for dry, store-cupboard staples – grains, flour, nuts, oils, and so on – not only less plastic packaging by a mile, but it could make you some savings, as buying loose items means you're not paying for a fancy brand name on a packet of chick peas or self-raising flour.

LAUNDRY AND HOUSEHOLD CLEANING REFILLS

To get started, invest in the largest-size containers of eco-brand laundry, household cleaning fluid and washing-up liquid, fabric conditioner and shower wash, for example – Ecover and Faith in Nature are two brands with a long tradition in refills, so you could buy large containers of their eco cleaning or toiletry products. Then, when you need to refill, locate your nearest health food or natural store that provides a refill service for these or similar products and brands. There's been a recent resurgence in stores offering refill shopping, and this is great news. The Whole Foods Market chain, for example, offers refills on Ecover.

Check for your nearest refill shopping options at http://plasticisrubbish.com/2015/08/16/refill-stores/

JUST ADD WATER

Alternatively, consider a switch to household cleaning products in concentrated form – soaps and cleaners, including laundry liquids and detergents. These are supplied as small sachets of active ingredients, which you simply mix with tap water and decant into refillable, reusable polypropylene cleaning bottles with trigger sprays.

The 'refill entrepreneurs' behind these new products are smart: not only are they doing their bit to help the planet and reduce plastics, but they are also aware that it's a waste of resources and money to transport weighty bottles of conventional cleaning solutions that are mostly made of water – which the customer can simply add at home.

Check out Splosh.com to start you off with a set of refillable bottles and concentrates.

JOIN THE BULK-BUY REVOLUTION: ZERO-WASTE SHOPPING

When I do my shopping now – wherever possible from small, independent shops – I'm always keen to know whether the retailer plans to embrace zero waste. Most express ambition, but are worried about the practicalities and the expense. Granted, it's not something that can be done overnight: it takes a lot of time and thought to de-package different lines and to look after perishable stock, especially foods and cosmetics. Our job as shoppers and plastic-reducers is to prove to small businesses and the larger retailers alike that we are the market, and that the market is there. That means showing commitment to zero waste by buying into refilling. If you're lucky enough to have a zero-waste shop near you – go!

Happily, though, over the last year the groundswell of support for turning the tide on plastic has led to some pretty cool new businesses devoted to unpackaged goods which you can buy in bulk. It took me several months to visit the latest branch of the Australian chain Source, in west London. It was a revelation.

These next-generation zero-waste stores are worlds away from the old-fashioned, sack-on-the-floor health food shops where you had to scoop your pulses and flours (and the odd weavil) into bags, which was all pretty uninspiring. From the machine that produces freshly ground nut butters (without added salts, sugar and fat), to the hoppers full of dried goods, from pasta to pine nuts; to the mini grain mill, the kombucha and different sorts of maple syrup on tap . . . My visit to Source not only replenished my food cupboard, but was an inspiration. It was also a tremendous relief not to have shop in fight mode, permanently alert to and disgruntled by plastic being foisted on me.

Many of the small, independent health food and organic grocery stores will sell both fresh and dry, store-cupboard foods loose and by weight. Whole Foods Market stores have a wide range of dry products to buy in bulk. Check online for local stores offering loose products:

- http://plasticisrubbish.com/2015/08/16/refill-stores/
- https://thezerowaster.com/zero-waste-near-you/
- thezerowasteshop.co.uk/
- https://www.beunpackaged.com/

And a number of online retailers are beginning to offer bulk home delivery services:

- Zero Waste Club, Organic Plastic Free Grocery https://www.zero-waste-club.com/

Bring your own containers, or pack in paper bags to bring home and decant into jars. On this note, a tip on kitchen kit – I recommend that you have a funnel and a ladle on hand for this task!

Bulk-buy the following and save on packaging: rice, grains, flours, pasta, soup mixes, beans, cereals, nut butters, sugars, dried fruits, nuts and seeds, snacks and treats. Bulk-buy loose teas and coffee, too, also herbs, spices, salts and peppercorns.

13

RETHINK

In this chapter, we'll tackle some of the habits, behaviours and special occasions that need a rethink in order to unlink them from plastic. We also need to rethink our approach when we're relying on organisations or venues outside of our homes and normal routines. It would all be so easy if we stayed inside, cocooned at home, planning for the next shopping trip or recycling day. The more bland, predictable and robotic your lifestyle, the easier it is to eliminate a lot of plastic. But who wants to live like that? There are high days and holy days, holidays and adventures to pursue, and these too can be approached as opportunities to cut out the plastic.

BREAK BAD HABITS

Put two bins in your bathroom I'm not talking wheelie bins – for most of us, space is at a premium in the bathroom, and if you're really short on space, try a small bin with two compartments. Label one for recyclables: loo roll tubes, empty bottles and tubs; the other for non-recyclables: wipes and cotton buds (with and without plastic stems), contact lenses – all of which in the UK are habitually flushed down the loo. Why we are even tempted to do this, I will never know. Wipes – which I repeat, contain plastic! – are the principle cause of emergency call-outs to plumbers, as they can easily get caught and block the pipe. Flushing them is playing Russian roulette with your loo. Why, why, why? Perhaps it's because we are not frank enough about this: Australians have no qualms about educating their kids that only the 3Ps – poo, pee and paper (loo paper) – should ever go in the toilet. If you really want to reinforce this message, write a sign and hang it above the loo. Marine experts consider anything other than the 3Ps to be litter.

Wean yourself off the wet wipe This is one addiction we need to rethink as a matter of urgency. The degradability of wet wipes is hotly contested: for years, the personal care giants have insisted that they are strictly tested for flushability. Countless tests by campaign groups have shown that although they can start to break up in the loo after flushing, wipes don't successfully disintegrate to the same degree as loo paper. This is hardly surprising because they contain plastic polymers which, along with plastic cotton buds and contact lenses (which also contain plastic) cannot always be recovered in waste-water treatment plants. Instead they escape into our waterways with devastating consequences for the environment.

Conservationists in London found just under 5,500 wet wipes that had amassed and congealed to form a bank in the Thames riverbed.[75] When the shape of the riverbed of one of the great rivers of the world is being altered by the thousands of throwaway wet wipes that we use for mere seconds, we really have to rethink our priorities. Change to a good old-fashioned flannel, use cut squares of muslin or reusable cotton pads that can be washed.

Buy fully compostable teabags Forty per cent of UK households keep a compost heap. This is potentially good news. Composting is a great, transformative way of reusing some of your food waste at home, so I thoroughly approve. Teabags, however, surprisingly, are unsuitable for home composting. Almost all the mainstream tea brands use a plastic sealant, polypropylene, to stop the bags from disintegrating in your teapot – they won't break down in the compost either, therefore, so will contaminate it. Seek out a fully compostable teabag or switch to tea leaves. Teapigs (teapigs.co.uk) call their pyramid-shaped bags 'tea temples'. They're made from a natural carbohydrate, cornstarch. The label is made from paper and even the ink on the label is vegetable-based. This means they are fully biodegradable.

Check garment labels I've briefly mentioned the emerging problem of microfibre pollution. Every day, thousands of microfibres from certain synthetic fabrics escape into the

75. Van der Zee, B.: 'More than 5,000 wet wipes found in an area next to the Thames the size of half a tennis court', *The Guardian*, 2 May 2018.

waterways from a single cycle in the washing machine. The garment manufacturers, clothing brands, policymakers and scientists need to come up with some smart solutions very soon, but already the state of California is working on legislation concerning washing and care instructions on clothing labels.

Potentially, garments that contain 50 per cent and over of synthetic fibres will have to carry a warning label. Consumers would therefore be alerted to the potential harm of microfibres and would follow a recommendation to hand-wash garments – without the agitation of a machine-wash cycle the garments shed fewer fibres.

One of the reasons we're attracted to synthetic fibres is because they're low-maintenance, so few of us will relish the prospect of having to start washing them by hand! Fleece fibre – sometimes made from recycled bottles – is known to be a heavy shedder as the fibres are quick to loosen. (The fleece is also, in my book, a fashion disaster as well as an ecological one! If there's a silver lining to the microfibre cloud, could it be the end of the fleece?)

So think before you buy, and invest in natural fibres instead. Wool, cotton, hemp and linen have all lost market share to synthetics in recent years. It is time to buck the trend – for the sake of the oceans.

LOOK OUT FOR HIDDEN PLASTICS

Wet wipes are a good example of a 'hidden' plastic. Many people don't even realise they contain polymers. The same can be said of single-use coffee cups, that we've talked about elsewhere. But the ones that shock me most include chewing gum (yes, this contains polymer); aluminium cans (a thin layer of plastic around the rim is used to prevent the tin from corroding); and teabags (see page 191). Knowing that these items contain plastic, we need to rethink our relationship with them and how we dispose of them.

RETHINK POLIC[illegible]

Lobby the airlines Some ai[illegible] issue of plastics and are r[illegible] on experiences. Although to a cli[illegible] is a secondary issue, given that [illegible]ying in the first place and generating g[illegible] emissions.

RETHINK THE HOLIDAYS

Use a bucket and spade library Like his fellow Margate resident Daniel Webb, Dan Thompson is an energetic plastic warrior – there must be something in the air down there! There is definitely something in the water: plastic, lots of it. When Dan Thompson spotted a plethora of abandoned buckets and spades in the town following a sunny bank holiday, instead of getting angry he made a large wooden box and decanted the beach toys into them. Now visitors to the seaside can borrow a bucket and spade from the Bucket Box at The Bus Cafe (thebuscafe.co.uk), instead of buying new from Poundland. It's a local solution we can take inspiration from.

PLASTIC-FREE FESTIVITIES

Throw plastic-free parties Celebrations have become heavily associated with plastic. A plastic-free party may, on the surface, sound like party-pooping: view it instead as a chance to have fun and get creative.

Call time on plastic cups and throwaway glasses, and those brightly decorated, so-called 'paper' cups and party plates, which are most often coated in plastic. Trust your guests with real glasses and china plates, or at least reusable tumblers and plates made of a durable heavy-duty acrylic or melamine plastic. Although both of these come with drawbacks, they can at least have long, useful picnic lives and be reused for years.

Straws must be paper, unless your guests carry their own (FYI jeweller to the stars Stephen Webster sells a reusable silver straw for £140, if you're looking for a collectable that fits into the swankiest parties).

Balloons If you must use balloons, they must never be released into the sky. Deflated and deposited by the wind, these are pernicious forms of wildlife pollution. There are biodegradable balloons on the market, but avoid these too. Some boast that they biodegrade as quickly as an oak leaf. But oak leaves don't biodegrade particularly quickly; it can take a few years, and in that time wildfowl or other species could easily have ingested the balloon. If balloons are to be used, make sure they are burst at the end of the night (a job for anyone who is a natural Grinch) and disposed of in the rubbish – they will be landfilled or incinerated.

Glitter is also a children's party flashpoint. Explaining to a four-year-old why it must be confiscated is not easy. If you do find yourself in this situation, feel free to borrow my strategy and explain that glitter is a ready-made microplastic and that penguins and other creatures might mistake it for food. Most children love penguins even more than glitter. Fortunately, you can now buy Eco Glitter, biodegradable glitter that microorganisms metabolise in the sewage system. Thank God, we can still shimmer.

Swap wrapping paper for a sustainable gift wrap Wrapping paper is another 'hidden' plastic: looks like paper, feels like paper, but increasingly it is plastic-coated. Wragwrap.com is a reusable and recycled fabric gift wrap in loads of different sizes, prints and colourways.

Unwrap Easter by switching to Easter eggs in cardboard or tins Easter eggs have become synonymous with overwrapping and the excessive use of moulded plastic. The average Easter egg contains 22 g of plastic packaging (so they too push up your plastic profile).

RETHINK POLICIES

Lobby the airlines Some are working on the issue of plastics and are receptive to exchanges on experiences. Although to a climate campaigner this is a secondary issue, given that you shouldn't be flying in the first place and generating greenhouse gas emissions.

Free-from-plastic festivals, football matches and other live entertainment You will know that you've adjusted your thinking when you check a venue's policy on plastic before you book to see your favourite artist! Once you're on the plastic crusade, you really can't switch off. I don't mean this to sound harsh or threatening; more of a friendly warning.

Occasions when you're likely to experience a mass of single-use disposable items can be disconcerting. Concerts and football matches are a case in point. I used not to notice that everyone got through several single-use plastic pint glasses and that they were shovelled into waste trucks afterwards: now it seems ludicrous. If every person was holding a plastic spade and threw it down on the floor afterwards I'd be pretty angry. Why would I make an exception for a plastic beaker?

This bothers me so much now that I look for festivals that have signed up to kick out single-use plastics. Sixty of the largest independent festivals, including Bestival and Kendal Calling, have committed to getting #drasticonplastic. Most are starting by banning straws almost immediately, but will quickly shift to reusable plastic beakers instead of disposable cups. At half a century old, and arguably one of the most famous festivals in the world, Glastonbury has led the charge on plastic, vowing to kick out plastic water bottles from 2019.

If not all of the big event organisers go this far, a festival should at least have a radical recycling policy in place. This year's Barclaycard BST festival uses cups made of HDPE plastic, collected, sorted and recycled into plastic hoardings and dispensers. By next year the aim is to turn the festival's plastic waste back into cups – that's closed-loop recycling. The sort of thinking I can get behind!

14

RECYCLE

We have arrived at our last R. With a new clear-sightedness about plastics and their flow into your life, hopefully you will soon have a lighter bin and a clearer conscience. The first seven Rs all have one thing in common: they are all about ways to pre-cycle – to head off plastics at the pass. If you've already used some of the tips in the preceding chapters, you are less likely to have been left flummoxed at the recycling bin because you will have Refused the bit of packaging with multiple heat-sealed layers of plastic. But two old clichés apply here: nobody's perfect, and old habits die hard. If this is your first attempt at culling plastic in your life, you will almost certainly be left with some plastic to recycle.

I've been tough on recycling in this book, and that also implies being tough on myself. (After all, I partly define myself as being a good recycler – if it was socially permissible I'd probably put it on my CV.) As we already know, the recycling system across the UK is deeply flawed. But one thing we do have in our favour is that we have good access to recycling schemes, albeit of a variable and confusing nature, depending on location. The key here is to be the best and most effective recycler you can.

Here is my top tip for great recycling.

Relocate to Wales! If you really want to up your recycling rate, move to Wales. Seriously. England, Scotland and Northern Ireland do not appear among the top ten world leaders in recycling. Wales, on the contrary, ranks second in the global household recycling league table, narrowly beaten by Germany but ahead of South Korea. Wales manages a recycling rate of 63.8 per cent, by comparison to England's rather weedy 42.8 per cent. This is especially impressive when you consider that twenty years ago, the newly devolved nation had a recycling rate of just under 5 per cent. Wales is now on track to meet its 70 per cent recycling target by 2025. The rest of the UK will struggle to reach its 50 per cent by 2020 EU target (a few years ago we all thought we'd reach this easily!).

So how did Wales become a recycling great? It seems that targets and goals really are important (there's a lesson for us all here). While England scrapped local council and national recycling targets, Wales established statutory targets that local authorities must meet. Most have put kerbside recycling docks in place, and rather than in commingled collections where everything gets shoved into one green 'mixed' recycling bag, residents sort recycling at home into different boxes which they then put into the appropriate dock.

This system results in higher-quality recyclate – what our recycling becomes after collection, sorting and processing – and means that the local authority finds it easier to sell into the global market. Over the past five years in the UK as a whole, the quality of recyclate has dropped. And when we talk about China's ban on our waste, we are really talking about a ban on low-quality recyclate. Admittedly the kerbside system means more boxes and bins and a bit more complication for us as householders, but ultimately it's for the greater good.

In or out of Wales, embrace the fact you have many bins, rather than a commingled system. Not only will the authority get a higher price for the recyclate, but also more markets are willing to buy it.

I was interested to see for myself what was so special about Welsh recycling, so last year I took a trip to Conwy, the pretty market town on the north coast of Wales. It wasn't my first trip there; the town has a castle that we were very fond of as kids. In those days I wouldn't have toured the local bins and recycling centre but, hey, interests change!

Not everybody there was a fan of the council's waste scheme. Some families told me that recycling had become a real chore: they had multiple bins (including for food waste), while black bag collections had decreased. Rubbish that couldn't be recycled and destined for waste to energy incinerators or to go to landfill was now only being collected once every three weeks. Some felt there was a lot of stick and not much carrot – to my mind, this was because they hadn't seen the carrot.

At Conwy's recycling centre the policy was paying dividends. As plastic and cans were processed into bales, the quality of the recyclate was visible: the bales seemed to glint in the sunlight. This shouldn't just be a matter of pride for residents; eventually they should also benefit through lowered council tax and better public services. If a local council's recycling system is running well, there should be more money for other public services including schools, libraries and health centres.

TOP TEN RECYCLING TIPS

But, if a move to Wales is out of the question, how do the rest of us with our less-than-perfect local recycling systems aim for shiny, happy blocks of recyclate? Here are my top ten tips on how to be a great household recycler using your local authority system:

1. ***Lobby your local authority*** to change to a kerbside-sort collection system, where waste streams are separated.

2. ***Sort, sort and sort again*** Even if your local authority only offers commingled recycling, where you throw everything into one bag or box, don't let this fact encourage a drop in your recycling standards. Even if it goes in the same bag, make sure it's clean and dry, and be careful not to contaminate the contents with any bits of plastic or lids that you're not sure about.

3 ***Do rinse out your plastics*** This makes your recyclables more pleasant and hygienic to hold on to until collection day. Also nobody wants to share their homestead with festering milk inside an HDPE pot, especially in summer.

4 ***Avoid Russian doll recycling*** By that I mean stuffing plastics inside other plastic containers. The main offending practice is putting straws and crisp packets into plastic bottles, and then replacing the cap. Increasingly MRFs (Materials Recovery Facility) are automated. Recycling is sorted by weight and air jets. By packing materials together, you confuse it, and your plastic will be rejected as contaminated.

5 ***Put the lids on bottles*** These are a different sort of plastic, but can be recognised in MRFs. Separated from bottles, lids are without a clear destination and are a total pain. They are also likely escapees if not attached to their bottle – the shape and buoyancy of lids mean that they are built for long voyages, and if they find their way into watercourses, they will travel and end up as an ocean pest.

6 ***Don't worry about peeling labels off your plastics*** Unless your council's instructions tell you to do so, this is unnecessary. Most waste is now separated at MRFs, where plastic that can be recycled is fed into a giant washing machine, and the heat and water will loosen the glue on labels and separate them.

7 ***If it's bottle-shaped and clear plastic, get it in the recycling*** A cursory check on plastic drinks bottles, by turning them upside down and checking for the PET sign, will tell you that these are eminently recyclable. If you are out and about, don't put empty plastic bottles in litter bins on the street – a number of local authorities do not recycle anything from litter bins (my council is one). Take them home and get them into your household collection. Otherwise it's a missed opportunity to recycle.

8 ***Never put items containing batteries in your recycling*** Lithium ion batteries, used widely in mobile phones, laptops, digital cameras and toothbrushes, are the biggest cause of waste industry fires.[76] Obviously, you should aim to eliminate these at source (see chapter 10, Refuse, page 155), but if you are throwing out old battery-powered plastic toothbrushes, they are technically Waste Electronic items, so drop them at your local Household Waste Centre to be processed correctly. You might also find drop-off points for used batteries in some supermarkets.

9 ***VHS videotapes are not welcome in recycling*** Yes, they are plastic, but the tape can stretch to 400 m in length and is coated in chemicals. On every visit I've ever made to a recycling centre, I have witnessed the havoc these can cause on the processing belts as the tape is pulled loose

76. *Flagstaff* magazine for the waste industry, Pennon, Winter 2017/18.

and wraps itself around every piece of equipment. The belt has to be stopped and recycling is delayed while the machinery is cut free. In chapter 15, Waiting for the Sea Change, I discuss the developments afoot in recycling technology that in the future will be able to deal with tricky items like videotape, straws and multi-layered plastic, but for now you need to store your VHS tapes until these options become more widely available.

10 ***Leave tricky moulded plastic of uncertain polymer, or those marked #3, out of your recycling.*** This could be polyvinyl chloride, or PVC, and cannot be recycled as it is a contaminant. I always try to buy incidentals such as torches and screwdrivers from mainstream retailers who have committed to phasing out PVC in their packaging. But from time to time this stuff gets into your house – typically purchased from a garage – do not let it get into your recycling.

OTHER RECYCLING OPPORTUNITIES

While I've made it clear that I think companies and brands should be made to take more responsibility for the plastic they put out on the market, some do offer forms of recycling. Make use of them!

Ocado runs a Bag Recycle Bonus Scheme. While this doesn't include the sealed packaging they use to deliver non-grocery items in their Fetch (pet products) and Sizzle (kitchen and dining ware) e-stores, you do get five pence for every plastic bag you return from your regular grocery shopping. You can hand back up to ninety-nine bags at a time, but don't make the poor driver wait while you collect them all: have them ready on your next delivery. Bags are returned to the warehouse to be sorted and recycled into more bags – a good example of the superior closed-loop recycling discussed in chapter 13, Rethink, on page 189.

Asda, Morrisons, Sainsbury's, the Co-op and Waitrose have collection points for plastic carrier bags and thin/flimsy plastic wrapping at their larger stores.

Tetra Pak has gone to town, and in nearly every town have established collection points for its packaging. Find your nearest collection point using their interactive map: http://www.tetrapakrecycling.co.uk/locator.asp

BUY RECYCLED

We're only going to close the loop if we buy recycled products too. Currently this is pretty difficult – fewer than 2 per cent of items on the market contain post-consumer plastic waste (i.e., the stuff we've already used that has been through the recycling system). One of the barriers to getting more recycled products on the shelves is thought to be us – the consumers! We have a predilection for buying clean, clear plastic – especially when it comes to food and drink. We all need to learn to love murky plastic, a sign it has had a second life, and can go on to have a third and a fourth.

At present there are not enough opportunities for us to benefit directly from recycling our empties. There are too few incentives for consumers. As waste plastic is increasingly seen as a resource, and designers and companies step up to the challenge of closing the loop and making more stuff from recycled content, I believe this will change. Bath and beauty brand Lush is ahead of the curve on this. Ninety per cent of its packaging is made from recycled material, including all Lush bottles and pots. Collect your empties and return them to store for recycling. What's in it for you? A free face mask!

Earlier I wrote that technically *anything* is recyclable. It is really a question of how much energy and money you are prepared to throw at the problem. Recycling company TerraCycle is testament to this fact. I first came across this innovative company that specialises in hard-to-recycle materials when I was reporting on pods for coffee machines. These are tricky little pieces of kit in terms of recycling: not only do you have an aluminium-foil pod, but you also have to grapple with the coffee grinds and a secretive thin film of plastic.

In 2015 the inventor of the original coffee pod, John Sylvan, said he regretted his invention because it was so unsustainable.[77] Recently things have improved. As the original coffee pods fell out of patent, compostable and refillable alternatives came on to the market. But throughout, TerraCycle remained unintimidated and collected the old pods for recycling.

77. Baer, D.: 'The Keurig K-Cup's Inventor Says He Feels Bad That He Made It – Here's Why', 3 March 2015, https://www.entrepreneur.com/article/243649

If you have a spare ten minutes, I urge you to spend it surfing the terracycle.co.uk website. It is much more fun than it sounds. The company's recycling repertoire ranges from the niche – disposable painting overalls (made from a plastic fibre, like face and cleaning wipes); large plastic exercise or birthing balls made from vinyl; backpacks; and baby food pouches. The list of possibilities for these recycling experts is seemingly endless. In fact, the TerraCycle strapline just happens to be 'recycle the unrecyclable'. You could, for example, fork out for an official cardboard recycling box, such as those for baby food pouches. Fill the box with your empties and when it's ready, phone for a UPS collection. If this sounds like an amazing way to target and deal with any stubborn renegades that evade the 8R strategy, be warned – it's going to cost you! Prices for a recycling box for baby food pouches start at £88.97, and for sports balls £97.20. This is definitely the deluxe route to turning the tide.

However, from time to time TerraCycle teams up with corporations or charities to tackle a particular plastic waste stream, including beauty product packaging, which is hard to recycle due to odd types of plastic, and empties that once contained cleaning products. These are much more affordable ways into their services. Keep an eye on the website, as a call will go out to register and become a collector/collection coordinator for a specific item, and to commit to collecting a certain amount. In my experience, however, these big recycling drives and other schemes funded by multinationals become oversubscribed very quickly.

It's a good idea generally to keep an eye out for charity

recycling initiatives that will ask you to send in or drop off particular plastic items. It's always worth checking directly with the charity to make sure it's legit. GHS Recycling (ghsrecycling.co.uk) based in Portsmouth are recyclers that partner with charities to collect plastic bottle tops from HDPE milk bottles. If you are planning your own collection on behalf of a charity, beware that they will only collect a minimum of 500 kg (which is a *lot* of bottle tops).

That's it! We have reached the end of our 8R steps to curb plastic consumption. If you've followed each of the steps in sequence, you will be at least four weeks through the programme, and at least two rubbish cycles in (if you're in Conwy in Wales, that's six weeks!).

Now is a good opportunity to repeat your plastic diary and check your progress. Again, fill in the grid, but this time only for a few days – unless you see no improvement, in which case something has gone terribly wrong! But I'm really hoping you'll notice a massive change in the amount of plastic in your bins. Hopefully, if you began with between eighty and 100 plastic items a week, you will have managed to halve that amount – at the very least.

The sense of satisfaction you'll get from this reduction should motivate you to a) carry on, and b) deepen your resolve. I hope you'll push on with your plastic reduction experiment until it's not an experiment, but a normal way of life. I believe this is part of an overall Better Living Strategy. I'm not saying you have to combine it with yoga or wearing only organic cotton (although I do support both of those activities), but you will be experiencing other supplementary gains – less waste in your house, less stress, perhaps some weight loss from fewer snacks – in addition to the kickback you get from engaging in a positive solution to plastic waste.

But enough about what's in it for you! In the next section, we'll address how to amplify your efforts and plug into the global movement that's going to reshape the way we view, interact with, use and reduce plastic in the near future and over the long term. We're going to turn the tide together, and make it count.

15

WAITING FOR THE SEA CHANGE

You can now say with confidence that you are doing your bit. Following a favourite maxim from the world of self-help, you are controlling the stuff that is within your control. You've instituted strategies and shopping lists, and perhaps even convinced reluctant family members to change their ways too. You've seen that small changes can yield big results. So now on to the next step. What big changes can you campaign for, cheer on and support that will help to bring about a decisive shift? There are a number of initiatives already in play that will bring about a sea of change. Adding our voice to the chorus calling for these changes will help turn the tide.

GROWN-UP POLICY

In 2007 the economist Nicholas Stern produced an environmental report, and while most people might never find the time to read it (at a hefty 700 pages, it wasn't exactly a summer reading essential) it helped to change the conversation on climate change. The Stern Review basically compared the cost of inaction to that of policy and regulatory changes. Guess what? It was substantially cheaper to act now and reduce emissions than to do nothing and wait for catastrophic climate change.[78]

Today, practically none of the manufacturers or retailers – the people who benefit from cheap, oil-based packaging – pay anything like the true environmental cost of cleaning it up. Moreover, the real cost of our collective plastic binge is yet to be properly accounted for. We need a version of the Stern Review on the cost of inaction on plastic. By April 2018 things were looking up when forty-two major businesses, including leading food and drinks brands, manufacturers and retailers, right through to plastic reprocessors, joined the UK Plastics Pact, convened by WRAP, and pledged to rethink the design, use and recycling of plastic. If not quite the radical move that campaigners had lobbied for, it was a step in the right direction, not least because the UK Plastics Pact spans the entire plastics supply chain.

78. http://webarchive.nationalarchives.gov.uk/+/http://www.hm-treasury.gov.uk/sternreview_index.htm

THE UBER COLLECTOR: BOTTLE DEPOSIT SCHEME

Campaigners on plastic had begun to sound like broken records, and the continuous refrain went something like this: 'Why don't we have a bottle deposit system in this country?' After all, the evidence from countries in the Baltic region with a deposit refund on beverage containers was compelling. In data from Finland, Germany, Estonia, Lithuania, Sweden and Denmark, plastic bottles do not feature in the top ten objects found on beach cleans. In the UK they edge into the top three.

In spring 2018 it was announced that after a government review, England would adopt the same system (Wales, Scotland and Northern Ireland have yet to make that decision). Although we still don't have a firm date for the scheme's introduction, food retailer Iceland has gone ahead and installed the first Reverse Vending Machine, part of the apparatus of the system, in its Fulham store to trial it. I am delighted by this progress. Bottle deposit systems boost recycling rates hugely. In Norway and Sweden, where such schemes have been running for a while, the collection rate for plastic bottles is a cool 97 per cent. In Norweigan, the process and system has its own word, 'Panteordning'. Perhaps that will catch on here, like 'Hygge' – the Danish word for cosiness and warm hearts and hearths – caught on, at least for a while.

It works by attaching a small levy to the drinks bottle:

across Europe it ranges between eight and twenty-two pence. The charge stays with the bottle as it moves from wholesaler to shop to consumer. As a customer, you pay the levy when you buy the drink, but when you return it you collect your money (normally via a voucher). The bottles are typically returned by popping them into a Reverse Vending Machine such as the one installed by Iceland. The machine scans the barcode and issues a refund voucher.

THE MATERIAL INNOVATORS

Brilliant innovations often arise incrementally, through years of trial, error and great effort. Just as the early inventors of plastic carried out thousands of experiments to create the first crude polymers, the innovators of today are jumping through hoops to find ways of reimagining them.

Nearly a decade ago, I came across Anna Bullus, a student of product design from Brighton. Her material of choice was unlike that of any other student I'd met before (and I have met many, and even filled in as a guest lecturer from time to time). Anna is obsessed with pre-chewed, second-hand chewing gum. If you had to pick a truly ghastly material to work with, this would fulfil most people's idea of the worst. But from the mouths of slack-jawed chewers, Anna saw potential. More precisely she saw that chewing gum contains synthetic rubber, a form of plastic that, if she could just get hold of it, could be recovered and put to good use elsewhere.

Like any good product designer, Anna is a solution-oriented person. She observed that the discarded, chewed gum was tough to remove from pavements without a lot of effort and a very powerful jet hose. As you chew, you swallow the nice bit of the chewing gum – the flavour and the sweetener – and you are left with what is essentially super-flexible plastic. If dropped on the ground, or stuck under a desk, these large hydrocarbons are squished out to become virtually inseparable from the surface.

Bullus devised her ingenious Gumdrop bin. In unmissable hot pink, the cute round bin seemed to entice people to deposit their gum in it rather than on the floor. If she'd have stopped there, that would have been a win, but the real genius of her innovation was in the material she used for the bin. It's made of Gum-tec®, her own polymer, which contains at least 20 per cent recycled chewing gum in the mix. Ten years on from our first meeting, there are hundreds of Gumdrops across the country, especially in areas with heavy student traffic (apparently big chewers) and at stations. When the Gumdrop is full – each can take 500 pieces – the whole bin is collected and processed as one piece of plastic.

Chopped up into flakes and then pelletised, the new plastic is sent to a factory in Leicester where the whole process begins again. The actual Gum-tec® formula is secret, but Bullus confirms that one full Gumdrop makes three more new bins. Since our first meeting, Bullus has expanded her range of products to include stationery, wellies, even the soles of shoes and coffee cups – all in her trademark hot pink.

Some might be squeamish about drinking from a coffee

cup made from gum that was once chewed by someone else, but I wish we would think less about the ickiness and our own preoccupations and celebrate the ingenuity. When it comes to reclaiming and reprocessing plastic, high-temperature processing destroys any trace of former use. What we have is a brand-new, eminently sterile material, ready for another life.

THE REIMAGINERS

The mindset of reimagination is crucial if we're ever going to embrace the circular economy. As we've seen over the course of these pages, we currently have a very linear approach to plastic. We buy it, use it, get rid of it. In many cases, the actual plastic is a side issue, considered worthless, if we stop to consider it at all. In the circular economy all materials have a value, especially plastic. They are not only reclaimed, but endlessly recycled so that, in effect, waste is designed out of the economy. Imagine if every company, every brand, every manufacturer was begging you to send their empties back. Imagine a tussle at your back door every time you were about to use the last drop of washing-up liquid – a scenario reminiscent of years gone by, when kids desperately waited to make Tracy Island as seen on Blue Peter.

This is, admittedly, a rather extreme imagining of what a future fully circular model might hold in store. But if it sounds fanciful, it is an idea that is gaining traction. In 2016 the Scottish Government announced a strategy that would move the nation closer to a circular economy. The strategy lays out a set of goals and actions that the Scottish government will undertake, including ambitious levels of household recycling and reprocessing, in order to build in circularity so that materials retain their value and are kept in use for as long as possible. Let's hope the rest of the UK doesn't get left behind.

The high priestess of the global circular economy movement is Dame Ellen MacArthur, something of a British heroine. Even if you didn't know much about the ocean or sailing on it, by the noughties you knew of Ellen MacArthur. Ellen brought sailing to a whole new generation when she took part in the Vendée Globe solo round-the-world yacht race in 2001 when she was just twenty-four. Finishing in ninety-four days, she became the youngest competitor ever to finish the race, the fastest yachtswoman around the globe and only the second person to go round the world solo in 100 days. It was one of those occasions when you felt like the entire nation was holding its breath, watching and agonising as she sent back her video dispatches. One thing we certainly got: this was a tough sport and she had what they call true grit.

Ellen MacArthur became one of those rare people who has experienced rationing resources and reusing everything around her – not as a whimsical experiment, but as a matter of life or death. In essence, her boat became a survival capsule that she endured for the ninety-four days of the race. She was acutely aware of her resources before, during and after those days. Everything was meticulously planned – even the amount of kitchen roll she could carry. Subsequently she seems to have seen the boat as a microcosm: it was her world full of supplies that had to be perfectly and exactly regulated so that she didn't run out. Unsurprisingly, it changed her perspective on the world and its resources in a profound and really useful way.

From this vantage point, and now retired from sailing, Ellen MacArthur is dedicated to tackling plastic and pursuing a circular economy. Through her eponymous foundation, she and her team have produced a blueprint for transformation: the New Plastics Economy – it's one of my favourites! What appeals to me is the vision it offers of a global economy in which plastics never become waste, and of how we can achieve that goal, step by step. In the same way that we can disrupt the flow of plastic into our homes by following some of the steps outlined in the previous chapters, the proposition outlined in MacArthur's blueprint disrupts plastic on a global and industrial scale. I'm clearly not the only fan: forty of the world's biggest brands have signed up to the New Plastics Economy. This includes the giant Unilever group of companies, from whom, statistically speaking, you most likely buy a big proportion of your household items. By 2025, all of the corporation's plastic packaging will be fully reusable, recyclable or compostable. I have to admit, it sounds great. If Unilever and others were able to achieve this, it could make our new, plastic-reducing lives problem-free.

MELTING AWAY

In the scenario above it would almost be as if our troubles with plastic might melt away like lemon drops. But I can't help picking up on the word 'compostable'. It's the most radical scenario, but is it even possible? Compostable or biodegradable plastics promise much, but will they deliver?

The UK is not only the birthplace of conventional plastic, tracking its heritage back to the nineteenth century, but it is also where some of the first defined oil-free bioplastics were created. In these, hydrocarbons are substituted for cellulose, the main substance in plant cell walls and the most abundant organic matter on earth. These bioplastics would – according to the theory – biodegrade in the earth as oxygen breaks down the chains of molecules which return to the soil without harm.

As early as the 1970s, some engineers and scientists cottoned on to the fact that increasing plastic production and usage each year was a very bad idea. Instead, the answer was to invent an alternative where you could control the degradation of the material, rather than having it hanging around for aeons. The late Professor Gerald Scott was an early champion of biopolymer research at Aston University in Birmingham, spearheading work on a biodegradable plastic that would not be made from oil. In 1972 he warned a journalist from the *Daily Mirror* that already that year the UK had got through 250,000 tons of indestructible plastic waste that would hang around for centuries: 'I would like to see artificial material made so that it can be reconstituted – used over and over again,' he said, sounding very much like Ellen MacArthur today. To that end, he developed a bioplastic.

For the early pioneers of bioplastics, timing the point at which the material would begin to degrade or dissolve proved to be a sticky issue, sometimes rather literally. If you think about it, this is a major concern. One of the bonuses of regular plastic is that you know where you are with it – it's virtually indestructible. But how do you make sure the polymers don't start to degrade while you're drinking a hot drink from a biodegradable cup, or running down the road in your synthetic sportswear? Too-rapid degradation could lead to embarrassment and, worse still, injury.

In Professor Scott's early experiments with the bioplastic material, the colour would change to an alarmingly violent hue, indicating that it was about to dissolve. It did not exactly catch on, although years later I would 'road test' a skirt and top by a major fashion label made from Ingeo – a bio-starch resin made from corn. This was a proposed solution to our increasing tendency to consume and discard fashion very quickly; at the time it was calculated that two million tonnes of clothing and textile waste hit landfill dumps every year. Rather than suggesting to spoiled consumers that we should consume less, the idea was that we could simply compost our clothes after we'd fallen out of love with them. Thankfully my outfit did not violently change colour or dissolve in the late afternoon, but it wasn't exactly a fashion win.

The designer Helen Storey, working with scientist Tony Ryan, has also designed pieces made from sugar. I came across a particularly beautiful dress in an exhibition. But you'll have to take my word for that, since at the end of the evening, it was dunked in water and dissolved before our eyes. Ingeo is now primarily used to make compostable food packaging.

While bioplastics might seem quirky, experts suggest we should get set for an 'aggressive explosion' of them. It is predicted that by 2050, nearly 50 per cent of the plastics we use will be derived from plants. This has led to worries about the use of arable land, needed for food production, being annexed for packaging. A short version of that quandary is, 'Would you rather feed the world or wrap a cucumber?'

But since the great awakening and our *Blue Planet II* moment, bioplastics are back in the frame. Up and down the country, I have met communities of anti-plastic warriors who are swapping out petro-plastics and switching to coffee cups and disposable nappies made with bioplastic. This strategy has the added attraction that it seems to offer a way to carry on using single-use disposable items.

But beware! I've met many a keen plastic-reducer who has been lulled into a false sense of security by bioplastics. They may be unlinking their consumption from oil, but it's not quite the full story. Bioplastics are still so niche that they present a bit of a problem: what's missing here is an end game. Some look and behave like regular plastic, so can easily end up in recycling. But made of plant plastic, they have a lower melting point and can cause havoc should they infiltrate the high-temperature cleaning process devised for their petrol-based cousins. Many people assume that they will degrade in landfill. But in landfill there is almost a complete lack of oxygen, which is needed to cause the chain of molecules to break down, and therefore, for the item to biodegrade.

Of course, bioplastic is known as 'compostable' for a reason. It should surely end up on a compost heap where it can break down into rich, dark organic matter and then help out in the garden. This also sounds like great news, especially given that 40 per cent of garden owners in the UK keep a working compost heap. But there's a snag here, too. In our temperate climatic zone, our compost heaps and bins only manage to reach typically weedy temperatures. There are probably just a couple of days in high summer each year when our compost gets close to the temperature required to kick off the breakdown of most bioplastic. A dear friend whose green values compelled her to use compostable nappies for her eldest son tells me she still comes across them from time to time when digging her garden. The nappies are pretty much intact. Her son is fourteen. That's what we term an eco fail, albeit a well-meaning one.

But if there's one thing we've been taught in recent years, it's that things are progressing quickly. It would be foolish to write this technology off just yet. Just as solar panels and electric vehicles have taken giant leaps forward to become credible alternatives, why shouldn't bioplastics have their day in the sun? From cutlery and plates to single-use coffee cups and lids and even face wipes, it is getting easier to lay your hands on products that declare themselves compostable. In truth there are a number of different symbols in use, which is confusing. Most suggest some leaves and an arrow indicating the leaves come from the ground. They should display the British standard, BS EN 13432, that tells you packaging is 'compostable', or BS EN 14995 for a 'compostable' plastic item. If you know the name of the company, you can type it into the database of one of the biggest global certifiers of bioplastic items to see if it's listed (https://www.bpiworld.org/CertifiedCompostable). But if you are in the market for bioplastic alternatives, think about collection. Some forward-thinking cafes and restaurants are signed up to compost collection schemes. Their bioplastic empties are spirited away to industrial composting facilities (known as In-Vessel Composters) that easily achieve the necessary temperatures.

Meanwhile, behind the scenes a chemistry revolution is taking place as innovators race to create plant-based polymers that will both perform as we need and biodegrade when we want. Niall Dunne is CEO of Polymateria, based in laboratories at Imperial College London. In the labs here, scientists are in pursuit of full biodegradability. This is clever stuff, including driving down the molecular weight of bioplastics to create molecules small enough for bacteria to break down efficiently. Dunne's stated aim is to create the 'Tesla of plastics', eliminating plastic pollution by radical innovation and doing for bioplastics what Elon Musk has done for electric vehicles, i.e., remove the stigma and make them highly desirable.

WEARING WASTE

Making plastic waste aspirational is key to the transformation of our relationship with it. We are simply that type of species. The more we covet something, the more we engage with it, and this hasn't been lost on sportswear giant Adidas. In 2014, Adidas produced its first collection of trainers made from ocean plastic – waste collected from the ocean and turned into high-tech yarn. Featuring a blue and aquamarine colour palette, the shoes looked to all the world like a very cool trainer, but the backstory and idea that we could help reduce ocean pollution made them incredibly sought after. Several collections later, Adidas has refined the marine-waste yarn until it resembles a high-tech cord. Consumer appetite for ocean plastic products, including trainers, continues to soar.

The Adidas range is the brainchild of Cyrill Gutsch, founder of Parley for the Oceans, an organisation that focuses the brains of creatives and designers on the problem of the plastic pandemic and asks them to find cool solutions. A former designer, Gutsch is evangelical about the role of consumers in the plastic pandemic. He thinks it's down to the modern-day consumer to save the oceans by buying smarter. Oh, and he considers oil-based plastics to be an epic design fail: 'It is time to do better and invent a smarter material.'[79]

79. Author interview and Engle, E.: 'Parley's Cyrill Gutsch on Why Reduce, Reuse, Recycle Is Invalid & How to Stop Designing with Plastic', *Core77*, 22 Feburary 2018.

FISHING FOR PLASTIC

This season, it's not only trainers that come in ocean plastic, either. From board shorts to evening dresses, the fashion world has embraced yarn made from recovered plastics. In September 2017, I attended the Green Carpet Fashion Awards in Milan, where A-listers strutted their stuff on a green carpet woven from Econyl, a recovered yarn made from fishing nets. The appetite is certainly there, but what of the practicality of recovering plastic trash from the ocean? And how much, realistically, can we remove? Given that there is estimated to be 80,000 tonnes of plastic floating in the Great Pacific Garbage Patch alone, surely it is a tall order to dream of removing even a fraction of it?

But this in essence is absolutely the dream of Boyan Slat, the Dutch wunderkind who first encountered the plastic pandemic when he was on holiday in Greece. Diving, he came across more plastic bags than fish, and started coming up with solutions. Based in Delft in the Netherlands, he dropped out of university and a degree in Aeronautical Engineering at the age of seventeen to create The Ocean Cleanup. Boyan's big idea – and everything he does is at scale – is to allow the ocean to clean itself. Inspired by the manta ray fish, he created a platform with long strings of floating booms that sift plastics from the water. The booms are connected to the ocean floor in a zigzag formation, allowing the ocean currents to run into them, trapping the maximum amount of plastic. 'The oceanic currents creating these gyres are not an obstacle – they are the solution. Let's use our enemy to our advantage,' he has said.[80]

In the summer of 2018 Boyan, now twenty-three, and his Ocean Cleanup team transported their mach II array from the workshop in Delft to San Francisco, where it was towed miles offshore. In underwater photographs, the array technology resembles a long string of giant air cushions being dragged behind a boat. After more testing, Boyan hopes to set the array to work on the Great Pacific Garbage Patch, trying to clean up as much of that 80,000 tonnes as possible.

Critics of Boyan – and there are plenty of them – allege that his ocean cleaning is a preoccupation and that both resources and his talent would be better spent if he focused on reducing our use of plastic. When he's interviewed on this point he tends to be sanguine, admitting that he doesn't know if he will succeed. 'I just thought it was important to at least try,' he told a journalist in 2017.[81] I admire that sentiment.

80. Slat, B.: 'How The Oceans Can Clean Themselves', by Boyan Slat, TEDxDelft, 2012. https://www.tedxdelft.nl/2012/10/video-how-the-oceans-can-clean-themselve-boyan-slat-at-tedxdelft/

Boyan Slat transcript: https://singjupost.com/oceans-can-clean-boyan-slat-tedxdelft-transcript/?print=print

81. Kelly, G.: '"It's important to at least try": Can this 23-year-old clear the oceans of plastic waste?', *The Telegraph*, 4 August 2017.

16

SQUAD GOALS

With the right innovation in alternatives to plastic, the right action and the right pressure on the right people in the right places, there is a very real chance that we can be bring the Plastic Age to an end.

This is the prize we have in our sights: talk about a golden opportunity. Granted, we're not quite there yet – we haven't entirely figured out what the ubiquitous material will be replaced with, and there are details 'to be ironed out', as they say. But we now have global consensus on the price of everyday plastic in the marine environment and that cost is just too great. We know that ocean plastic waste has catastrophic implications for sea life and for the food chain and can act as a vector for some of the most pernicious chemicals ever created. On 6 December 2017, all 193 UN member states signed a resolution to eliminate plastic in the sea. The signatory states resolved to monitor how much plastic they were dumping in the ocean and to explore ways of making the practice illegal.

In the UK, we also have an abundance of new legislation, targets and goals and a job to do, as a result. Everybody – and by that I don't just mean us, the consumers, I mean every public body in the UK – has a plastic reduction target, from Buckingham Palace and Parliament, to the big corporates. These targets bring together a timetable of actions such as stopping the use of plastic straws (an entry level pledge), the promised introduction of the Bottle Deposit Scheme and the possible extermination of pests such as wet wipes.

TURNING UP THE HEAT

So, while we have reasons to be cheerful, some of the target goals are way off in the future. The government's own environmental targets are set for 2042, by which time billions of tonnes of new plastic will have been created. So how do we turn these promises into action, add pressure for more to be done sooner, now, today?

This progress in international UN resolution and local legislation would simply not have been possible had the Plastic Age not run slap-bang up against the Activist Age. The activists are winning, and we need to keep it that way.

The power of collective action cannot be overstated. We must think of ourselves as global citizens and appreciate that we have agency and influence. Yes, we are consumers, but that's not all we are, and we shouldn't confine our ambitions just to buying better stuff; spending power is only one of the weapons we have at our disposal in the action on plastic. We have a lot more to give.

By tackling plastic at source in your own life, you've already made your mark in the global movement. But it doesn't have to stop there. There's a wealth of opportunity to amplify your action and add your voice, to get behind those campaigns and organisations that will apply the pressure where it is most effective.

When you're deciding which campaigns to get involved with and where to put your name, and indeed your energy, it's useful to have in mind the qualities that make for a successful movement. Look for campaigns and associations that are helping to collect data that's feeding into global data sets on single-use plastic and toxins in the world's oceans, and feeding back into the UN's goals. Look for those that are creating an online community of change-makers with global ambassadors who will push the message out.

One of my favourite action groups is A Plastic Planet (aplasticplanet.com), co-founded by entrepreneur Sian Sutherland. A Plastic Planet's campaign has a simple mission – to turn off the tap on plastics with four pillars of action: lobbying, education, media and industry.

Increasingly we're witnessing the power of coming together and pooling our agency. In the Netherlands in 2015, a group of over eight hundred citizens got together to sue the Dutch government on the grounds that it had knowingly contributed to a breach of the 2°C maximum target for global warming. The Hague ruled in their favour, ordering the Dutch government to take action to cut greenhouse gas emissions by 25 per cent within five years. It was the first time a court had ordered a state to protect its citizens from climate change.

The ruling represented a major victory for citizen action, but it was also a sign that activism was switching up a gear. James Thornton, CEO of environmental law organisation ClientEarth, described the victory as 'remarkable'. 'A major sophisticated European court has broken through a political and psychological threshold,' he said.

Could we, through group action, engineer a similar breakthrough on unnecessary plastic packaging, perhaps?

THE POWER OF SMALL AND LOCAL

I know only too well that time is a precious resource. You are probably already at full tilt, and can't necessarily dedicate extra hours in your day to campaigning on this issue. But there are ways to sign up without adding significantly to your workload.

Start by signing a pledge to #BePlasticWise at Ocean Wise http://pledge.ocean.org.

Adding your voice to this and similar global campaigns also helps to remind you that you are part of a greater mission. Never underestimate the power of a simple petition, either.

Wrexham gardener Mike Armitage was so perturbed when he discovered that the teabags made by PG Tips, Britain's biggest tea brand, which he had been putting on his compost heap, contained between 20 and 25 per cent plastic, he started a petition demanding change using the campaign website 38 Degrees, a UK not-for-profit political activism organisation. He quickly got 200,000 signatures, and sent it to Paul Polman, CEO of Unilever, who announced that they would design plastic out of their PG Tips teabags. So that's one man, one petition and a spectacular outcome: ten billion fewer plastic teabags damaging our environment every year.[82] Other teabag brands are now under pressure to follow suit.

WHAT TYPE OF ACTIVIST ARE YOU?

Many people are nervous about the term activist, or at least find it off-putting. Activism used to mean placard-waving and chaining yourself to fences, and there is still a place for this, especially when a specific site is being threatened.

But there is also a school of thought that suggests we should match our activism to our temperament. Sustainability expert Solitaire Townsend, herself the antithesis of the hair-shirt model, unwraps this idea in her book *The Happy Hero*. There's even a profiling section where you can match your personality more specifically to different forms of activism. It's a bit like dating for good causes, but the payoff is crucial. Townsend argues that if you do activism right, you'll get the most spectacular kickback: it will make you happy.

82. https://home.38degrees.org.uk/2018/02/28/pg-tips-won/

If taking action on a cause that matters deeply to you makes you happier, you are more likely to stick at it, and less likely to suffer from activist 'burnout', which you are probably going to feel if you are swept up in an angry movement that doesn't sit with your temperament.

How to be a Craftivist by Sarah Corbett has also helped me to reframe my idea of activism. As an introvert with very strong beliefs on climate change and social justice, Corbett found herself burned out from too many shouty protests. She realised she needed a different style of resistance, and embraced craftivism. A term coined by writer Betsy Greer, she describes it as 'a way of looking at life where voicing opinions through creativity makes your voice stronger and your compassion deeper'. Corbett began embroidering messages on hankies asking for workers in the garment industry to be paid a living wage. She inserted them into the pockets of shareholders at AGMs for high-street big-brand FTSE 100 companies. As a result of Corbett's quiet craftivism, the living wage was timetabled as an official motion.

I'm already starting to see creativity and pragmatism come together in fighting the plastic pandemic. There's resistance in making a simple bag from recycled fabric to keep your bread in when you buy it unpackaged from the bakery, or sewing a simple pouch to keep reusable cutlery in your bag. If you know a heavy and unrepentant user of plastic cutlery, gifting them a hand-sewn cutlery bag is a non-confrontational way of getting them to engage. Embroider their initials on it, if you really want to push the boat out!

ADVENTURERS

You might want to plug into an expedition or activity that adds an adrenaline rush to your plastic protest.

Lizzie Carr has pioneered the 'paddle against plastic'. She is a British adventurer who in 2016 become the first person to successfully stand-up paddle-board (SUP) the length of England along its connected waterways. She travelled 400 miles in twenty-two days, and plotted and mapped out plastic pollution along her route. Not only did she bring back a lot of evidence that increased our understanding of how plastic collects and travels in our waterways, but she has inspired hundreds of us to shakily get up on paddle boards. From here, the vantage point is particularly good when it comes to plastic-spotting, as you look directly out and over the water. Be prepared to build up some competency before you can start fishing out plastic!

What I like about the current crop of British adventurers and expeditioners is that they are also using their fight against plastic to raise the profile of women in science. Emily Penn is an ocean advocate and a veteran of epic voyages, who sets sail in summer 2018 on an expedition through the Great Pacific Garbage Patch (see page 78) with a twenty-four-woman crew. They will make daily trawls for plastics and pollutants, and collect data for global data sets and scientific studies.

Places on similar 3,000-nautical-mile voyages that could be described as 'challenging' are available at https://www.earth-changers.com/sustainable-places/pangaea-exploration.

Let's face it, they're not for everybody, but it is worth following expeditions like this through social media and live Facebook links. The technology and communication of the ocean science community has come on in leaps and bounds, and connecting with what's happening in the field (in this case, the place in the world where plastics have most densely accumulated) is an eye-opening reminder of the worldwide picture.

A PROBLEM SHARED IS A PROBLEM HALVED

You don't have to venture far to turbocharge your plastic activism. Plugging into your local community is arguably the most valuable thing you can do to turn the tide. Think about how much you have scaled down your own plastic consumption during the course of reading this book, and beyond. You will have removed hundreds of needless plastic items from your life. Now imagine how you might persuade every friend, relative and neighbour to do the same. When you get talking and campaigning in local shops and businesses, from hairdressers to greengrocers, the gains have the potential to grow even more. All of these premises have behind-the-scenes plastics in their supply chain, waiting to be tackled. Many people will share your concerns, but may not know where to start.

This is the principle behind the Plastic Free Communities programme run by Surfers Against Sewage: https://www.sas.org.uk/plastic-free-communities/ Communities develop an action plan, with advice from the Surfers Against Sewage local representative. Through a series of meetings between residents and business owners, they talk out the issues, share suppliers of alternative plastic-free or compostable products, and even consider the waste and collection facilities in their local area. Once they have tackled the major points and developed a plan of action, they are awarded a coveted Plastic Free Community status and certificate.

The goal is to have 125 certified Plastic Free Communities by 2020. They are well on their way.

I visited the first Plastic Free Community in Penzance in Cornwall on a cold winter weekend at the end of 2017 along with almost the entire population of the village of Aberporth in Wales. The visiting residents of Aberporth included interested householders and cafe, shop and pub owners, who had made the seven-hour journey to Penzance to learn what it took to achieve plastic-free status. Plastic Free Community leaders in Penzance welcomed us all and together we toured the fine town, from the chip shop on the front to the bustling cafes. We talked straws, litter, tourism (both towns are tourist hotspots), engagement and a lot about single-use coffee cups. We swapped stories, ideas and – most importantly for the business owners – contacts of suppliers of innovative products that would release them from the burden of single-use plastics. And you know what? It was fun. A highly recommended day out. Six weeks later, Aberporth was officially awarded its Plastic Free Community certificate.

Now it's time for your town, too.

EPILOGUE

Towards the end of my own plastic experiment, implementing the 8Rs and strategy outlined in these pages, I moved house. Aside from the issue of trying to avoid using rolls and rolls of bubble wrap – a 1957 invention originally intended for use as wallpaper, but by now a ubiquitous fixture in life, especially in the removals business – the move allowed me to up the ante on my plastic reduction drive.

Moving from a flat into a house, I was suddenly presented with extra shelves and cupboards that could hold loose, unpackaged produce. It also gave me an outside shed where empty reusable laundry and cleaner bottles could be stored. The zero-waste store that fortuitously opened as I began to write this book means that bulk-buying unpackaged goods and refilling containers has become a reality. Perhaps I'm on the road to joining the zero-waste elite on Instagram after all – although with fewer white minimal surfaces.

I also got closer to the water, fulfilling a long-held ambition.

My new house is slap, bang next to the River Thames. I've always dreamed of living next to this great river, and I've seen it from every angle. This includes the usual river trips, but over the years I've taken part in insect surveys along the banks and in river clean-ups. I've viewed it from the sky in a Lynx helicopter: in 2012 on a reporting assignment, we hovered above the Royal Navy's largest ship, HMS *Ocean* as she squeezed – and it really was a tight fit – through the Thames Barrier. It was the last time a ship of this size would sail up the Thames, and one of the last official outings for HMS *Ocean*, which afterwards was decommissioned and sold to Brazil.

I've always been acutely conscious that I am living in one of the great coastal cities. London is qualified by the huge estuaries that serve it, and by the volume of water that passes through and around it. Now that I'm actually living on the river's banks, it has reignited my activism. Today we are aware that the threat to the coastal ecosystem is not only from climate change, and its attendant rising seas, but also from plastic. So as well as joining my neighbourhood clean river groups, I've got my own net, so I can at least fish out the macro plastics that float near me, and I've treated myself to a kayak, so I can join the dozens of canoeists and kayakers who litter-pick on the waters of a weekend.

I've also noticed another change coming over me. Back at my old flat I had a long-running debate with my local council, who refused to give me access to kerbside recycling. Unlike the houses in our street, multi-tenancy buildings such as mine weren't permitted recycling bins. In the end I managed to procure a couple regardless, and a friendly neighbour used to let me plonk them outside his house on recycling day.

Now, in a house in a different borough, I am no longer bin-impoverished. In fact, I have an abundance of big bins – three wheelie bins and a composter, for starters – and a council that offers kerbside recycling (superior to commingled, as we know). But am I happy? Of course not! I have found myself joining the bin-whingers, because the pretty front garden is cluttered with them. After sulking about this for a while, I've realised it's in my power to regain control. By stepping up my plastic-free efforts, by applying the 8Rs with laser focus, I can free myself from all but the smallest amount of rubbish. My goal is to hand all the wheelie bins back to the council because I no longer need them, and therefore free up my garden.

If handing back wheelie bins constitutes personal success as far as I'm concerned, what of our collective ambition? What should that look like, and how will we know when we have done enough? To borrow from cringeworthy management-speak: What will success look like?

I received an answer to that back on the beach at Sennen Cove in Cornwall during the 2018 Big Spring Beach Clean. It's a tough clean at Sennen Cove: the plastic washed up by the dynamic surf tends to get jammed under the large stones and boulders that cover the back of the beach. The kids, tweens and teens that populate the beach all year around, and pitch in on the big beach cleans, are undaunted. Many are grommets – surf slang for young surfers, sometimes abbreviated to 'gromms' or 'gremmies' – and are here on a daily basis, at least in the spring and summer. They skip across the rocks like mountain goats in sparkly wellies and trainers. Even the little ones, who instinctively form little teams, lifting the rock clear while another goes in to extract the plastic like a beach surgeon. It's impressive.

Dave Muir, the owner of the beach's surf school and a Surfers Against Sewage rep is passionate about plastic. He holds a 'mini beach-clean philosophy'. Before any of his grommets enter the water, they must pick up a handful of plastic. Every little bit helps. As we watch tousle-haired kids dragging full bags of plastic waste off the beach, I tell him he has taught them well. By the end of the day, they'll have done such a good job at cleaning up that the next day's crop of beach cleaners will be forced around the headland to Gwenor in search of plastic.

Dave, who lives around the headland with his young family, agrees. But then he adds, 'Isn't it a shame, though, that they have to do it? When I was a kid, I just came down here, ran into the waves and surfed. I didn't have a care in the world.' It's an important point. My childhood was also free of edicts and directives telling me I must pick up three bits of plastic before I charged into the sea.

But it gives us something to aim for. The day when kids everywhere can return to a carefree state and enjoy the great outdoors without worrying about the plastic pandemic will be a signal that we've won. Throughout this book I've championed the power and importance of group endeavour epitomised by the beach clean. By joining together in constructive, concerted action, we all have to hope that there's a beach clean in our future where we struggle to gather even a single bag of waste. That's when we will know that we have truly turned the tide.

PLASTIC DIARY GRID

Plastic / product	Source	Status: **A** = *avoidable* **U** = *useful* **N** = *necessary*	Number of uses: **S** = *single-use* **M** = *multiple-use*	

	U = *uninvited* **P** = *purchase (or given with product)* **F** = *free*	**Features**	**End:** **B** = *general bin (landfill or incineration)* **R** = *recycling*	**Number**

FURTHER RESOURCES

We have an abundance not just of plastic, but of a more welcome resource: scientists, researchers, bloggers, campaigners and constructive citizens all doing something to halt and then reverse the pandemic. There is so much activity and energy that it's not possible to list all of the interesting and significant players here, so I've confined this list to my favourite resources. It runs from local agitators to global players and quirky citizen science projects. All of the sites are rich in content, and include links to evidence-based research. Many of these continue to feed into the global campaign for change on plastics and to important scientific data sets.

GLOBAL RESOURCES

5 Gyres Institute

www.5gyres.org

One of the original leaders and evidence-gatherers on the plastic pandemic, 5 Gyres Institute is a treasure trove of research and action. Also includes the campaign of 'plastics to ban 2.0'. Well-resourced and continually updated.

Ocean Recovery Alliance

www.oceanrecov.org

Also the home of the Global Bottle Deposit Challenge calling for bottle deposit systems worldwide, and the Plastic Disclosure Project that brings industry and brands together to account for plastic use (plasticdisclosure.org).

United Nations Environment Programme (UNEP)

www.unenvironment.org/explore-topics/oceans-seas

Nearly two hundred countries are signatories to a United Nations resolution to eliminate plastic in the sea. UNEP works to make this resolution effective. Also runs the #CleanSeas campaign that makes the science behind ocean plastics relatable and understandable.

CAMPAIGNING ORGANISATIONS

A Plastic Planet

aplasticplanet.com

The thinking behind UK-based A Plastic Planet (APP) is that we can't unknow about the plastic pandemic. Action and change are the only responses. A great resource for campaigning, APP runs campaigns across industry and the media to lobby and educate in order to keep the pressure on.

Greenpeace

www.greenpeace.org.uk/what-we-do/oceans/plastics/

Make your own plastic pledge, but also sign up to current petitions to apply pressure on government and brands. Current campaigns include requiring UK supermarkets to eliminate unnecessary and non-recyclable plastic packaging by 2019.

The Plastic Soup Foundation

plasticsoupfoundation.org

Through campaigns, news articles and social media, Amsterdam-based the Plastic Soup Foundation draws attention to the 'plastic soup' effect that plastic causes in the ocean. It is particularly strong on eye-catching, art-led installations and stunts such as the Plastic Soupermarket, a touring exhibition of fake supermarket items made from plastic waste.

For a Strawless Ocean

strawlessocean.org

Take inspiration from Seattle and an A-list roster of anti-straw campaigners who have made short (and often funny) films in an attempt to get straws banned. They've been successful. The Mayor of Seattle announced that in July 2018 Seattle will become the largest metropolitan city in the world to ban the single-use plastic straw.

Sky Ocean Rescue

skyoceanrescue.com

This media company has long-running campaigns on single-use plastics and is also the home of #PassOnPlastic for the football industry.

City to Sea

www.citytosea.org.uk

Bristol-based but with a national focus, this organisation is expert on waterways and coastlines, and menaces such as tampon applicators and cotton buds. Also runs a national campaign to promote using tap water on the go instead of single-use water bottles.

OneLessBottle

www.onelessbottle.org

Londoners are the biggest users of plastic water bottles per capita, and this is the leading campaign and resource to get them to cut down or stop.

PLASTICS AND RUBBISH

The Waste Resources Action Programme (WRAP)

www.wrap.org.uk/content/the-uk-plastics-pact

A mine of information on all materials and recycling, WRAP makes figures on our waste and consumption habits in the UK digestible. WRAP is also bringing industry and retailers together to transform plastic packaging as part of the UK Plastics Pact.

Zero Waste Scotland

zerowastescotland.org.uk

The home of Zero Waste Scotland that also supports innovation that will help remove single-use plastics. If you have a great idea, this is a good place to send it!

Recycle Now

www.recyclenow.com

Find out where you can recycle plastics near you with a postcode search.

SHOPPING AND LIFESTYLE

The Zero Waster

thezerowaster.com

Great guide to UK shops selling bulk and unpackaged goods.

plasticisrubbish

plasticisrubbish.com

Blog and resource for cutting plastic and living compostably.

Friends of the Earth

friendsoftheearth.uk/latest/plastics

Regular guides include how to have a plastic-free Ramadan and an analysis of plastic-free wet wipes.

Plastic Free Hackney

plasticfreehackney.com

Bettina Maidment is a mother of two young boys and blogs on bringing up plastic-free kids.

Responsible Travel

www.responsibletravel.com

The travel site that specialises in low-impact and ethical trips now has a 'no single-use plastics' holiday section. Trips, holidays and experiences are clearly labelled to make it easy for travellers to see where and how they are plastic-free. Also features a dedicated online guide to plastic-free travel, full of tips.

Soneva

Soneva.com

An innovator in the plastic-free movement, Soneva luxury resorts in Thailand and the Maldives exclude plastics. Even if you can't quite justify a trip, this is a great resource for those who want to see how a glamorous zero-tolerance approach can look. The company banned the use of plastic water bottles in 2008, so has plenty of experience.

SCIENCE AND TECH

The Plastic Tide

theplastictide.com

Become a citizen scientist here without even leaving the sofa. So many drone images of beaches all around the world, including the UK, have been taken that researchers need help finding the plastic on them. You can sign up to review the images and tag plastic waste on these beaches.

Pangea Exploration

panexplore.com

This is where you apply for a place on a scientific voyage aboard a seventy-two-foot ex-Global Challenge sailing yacht repurposed as an adventure research vessel.

Plastic Drift

plasticadrift.org

Use the map function here to find out how plastic travels in the ocean over time.

JAMSTEC Deep-sea Debris Database

www.godac.jamstec.go.jp/catalog/dsdebris/e/

The Japan Agency for Marine-Earth Science and Technology (JAMSTEC) has made their database of images from over five thousand dives by individuals and submersibles publicly accessible here. This is for people who like a data crunch: debris is classified by shape and material. You can also view videos and imagery of sunken plastic in deep-sea locations.

The Ocean Cleanup

theoceancleanup.com

Founded by Dutch inventor Boyan Slat, the organisation employs cutting-edge technology to remove plastic from the oceans. This is where you can find out how they build their super-arrays in Delft, Holland that claim to capture plastic, and follow them as they are deployed around the five oceanic gyres where plastic is building.

Polymateria

polymateria.com

Based at Imperial College London, Polymateria is engaged in the latest in biodegradable plastics research.

DESIGN AND BRANDS

The Ellen MacArthur Foundation

ellenmacarthurfoundation.org

A major driver in the shift to a circular economy and founder of the UK Plastics Pact, the EMF is a great resource for reports that map out a new future where plastic waste won't exist, such as the New Plastics Economy roadmap.

Parley for the Oceans

www.parley.tv

Brings together designers and creatives to rethink plastic waste in the most radical ways through collaborations with film-makers and brands.

Econyl

econyl.com

A company that takes plastic ocean waste and regenerates and spins it into yarn for clothing brands.

Ecover

ecover.com

The green cleaning brand has been rethinking plastic for over a decade. Its 100 per cent plant-based washing-up liquid bottles are now a fixture in any self-respecting eco household. They are just the tip of the iceberg in a 'clean plastic' innovation journey.

Finisterre

finisterre.com

Cornish surfwear brand that also campaigns to increase awareness of microplastics, and produces collections made from regenerated yarn from fishing and plastic ocean waste.

Guppy Friend

http://guppyfriend.com

Sells a laundry bag for washing synthetic apparel which claims to trap microfibres shed during the wash cycle. Good resource for information on microfibres.

IS THERE AN APP FOR THAT?

When I'm out and about, I've found it useful to have some key apps on my phone (all downloadable from iTunes). These are my top three:

Zero Waste Home

zerowastehome.com

Bea Johnson's Zero Waste Home app features hundreds of tips and shopping lists for cutting out plastic, and a map to find the best grocery stores dedicated to unpackaging.

Beat the Microbead

beatthemicrobead.org

In the UK we've banned microbeads in some products, but not all. It can be hard to determine if a product contains them. This is further exacerbated if you're travelling abroad, where there may not be any legislation. Download the Beat the Microbead app and you can scan many products before you buy, using the barcode, to detect whether they contain microplastic ingredients.

Giki

gikibadges.com

Giki rates 250,000 supermarket products on health, sustainability and fairness issues and increasingly addresses plastic packaging, too. The less plastic, the higher the rating. The app is in the process of developing a 'better plastics' badge.

Surfers Against Sewage

Sas.org.uk

The SAS website includes advice, tips, pledges, campaigns and resources such as the unmissable Surfers Against Sewage Plastic Free Coastlines Community Toolkit:

https://www.sas.org.uk/wp-content/uploads/Plastic-Free-Coastlines-Community-Toolkit.pdf

The organisation gets special mention here (and overleaf) because they are the engine of change in the UK where plastics are concerned, and I don't think we'd be having this conversation were it not for them!

SURFERS AGAINST SEWAGE

Surfers Against Sewage is a national marine conservation and campaigning charity that inspires, unites and empowers communities to take action to protect oceans, beaches, waves and wildlife.

Surfers Against Sewage has the support of thousands of members across the UK. Together, we speak out for the protection of the coastal environment – your oceans, waves and beaches.

We're not just surfers, and we're not just about sewage. We're a voice for all water users and coastal enthusiasts, from surfers to swimmers, canoeists to holidaymakers. Anyone who loves going to the beach but hates seeing it polluted – that's who we speak for.

Our HQ is in St Agnes, Cornwall, but we cover all 19,491 miles of UK coastline, protecting beaches, monitoring water quality, organising beach cleans, running educational tours, lobbying government and industry, reporting on pollution, and campaigning for the protection and conservation of your local spots.

WHAT WE DO

We Protect the Environment

We call for better legislation and stronger action to address complex environmental issues, including marine litter, sewage and diffuse pollution, climate change and coastal development. We also take direct action, creating a powerful network of coastal campaign leaders and running community beach cleans and awareness campaigns to target these issues head-on.

We Challenge Industry

We produce and promote scientific, economic and health evidence to support calls for a cleaner and safer marine environment. We also lobby industry to adopt better standards to protect our coastlines, marine life and seas.

We Influence Government

We talk to MPs, MEPs and county councils about key issues affecting oceans, wildlife, beaches and recreational water users, and the policies and legislation needed to better protect them.

We Motivate People

We create volunteering opportunities for individuals and communities to be proactive in safeguarding our seas, coastlines and beaches. Our community-led beach cleans remove tonnes of marine litter every year, and our education programme inspires thousands of school children nationwide to get involved. We also support communities with environmental initiatives on achievable, sustainable solutions, which can help protect our waves, oceans and beaches.

For more information on all of our campaigns, environmental initiatives and opportunities, please visit our website:

sas.org.uk

Registered charity Number: 1145877

Website: sas.org.uk
Twitter: @sascampaigns
Instagram: @surfersagainstsewage

INDEX